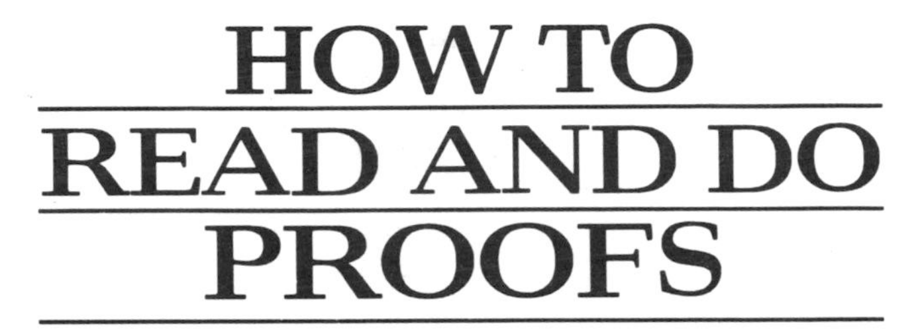

an introduction to mathematical thought processes

HOW TO READ AND DO PROOFS

an introduction to mathematical thought processes

SECOND EDITION

DANIEL SOLOW

Department of Operations Research
Weatherhead School of Management
Case Western Reserve University

JOHN WILEY & SONS
New York · Chichester · Brisbane · Toronto · Singapore

Library of Congress Cataloging in Publication Data:

Solow, Daniel

How to read and do proofs : an introduction to mathematical thought processes / Daniel Solow. – 2nd ed.

p. cm.

Includes bibliographical references.

ISBN 0-471-51004-1

1. Logic, Symbolic and mathematical. I. Title.

QA9.S577 1990 89-70592

511'.3--dc20 CIP

Printed in the United States of America

20 19 18 17 16 15 14 13

Printed and bound by Courier Companies, Inc.

To my late father, Anatole A. Solow,
and to my mother, Ruth Solow

FOREWORD

In a related article, "Teaching Mathematics with Proof Techniques," the author has written, "The inability to communicate proofs in an understandable manner has plagued students and teachers in all branches of mathematics." All of those who have had the experience of teaching mathematics and most of those who have had the experience of trying to learn it must surely agree that acquiring an understanding of what constitutes a sound mathematical proof is a major stumbling block for the student. Many students attempt to circumvent the obstacle by avoiding it—trusting to the indulgence of the examiner not to include any proofs on the test. This collusion between student and teacher may avoid some of the unpleasant consequences—for both student and teacher—of the student's lack of mastery, but it does not alter the fact that a key element in mathematics, arguably its most characteristic feature, has not entered the student's repertoire.

Dr. Solow believes that it is possible to teach the student to understand the nature of a proof by systematizing it. He argues his case cogently with a wealth of detail and example in this book, and I do not doubt that his ideas deserve attention, discussion and, above all, experimentation. One of his principal aims is to teach the student how to read the proofs offered in textbooks. These proofs are, to be sure, not presented in a systematic form. Thus, much attention is paid—particularly in the two appendices—to showing the reader how to recognize the standard ingredients of a mathematical argument in an informal presentation of a proof.

There is a valid analogy here with the role of the traditional algorithms in elementary arithmetic. It is important to acquire familiarity with them and to understand why they work and to what problems they could, in principle, be applied. But once all of this has been learned, one would not slavishly execute the algorithms in real-life situations (even in the absence of a calculator!). So, the author contends, it is with proofs. Understand and analyze their structure—and then you will be able to read and understand the more informal versions you will find in textbooks—and finally you will be able to create your own proofs. Dr. Solow is not claiming that mathematicians create their proofs by consciously and deliberately applying the "forward–backward method"; he is suggesting that we have a far better chance of teaching an appreciation of proofs by systematizing them, than by our present, rather haphazard, procedure based on the hope that the students can learn this difficult art by osmosis.

One must agree with Dr. Solow that, in this country, students begin to grapple with the ideas of mathematical proof far too late in their student careers—the appropriate stage to be initiated into these ideas is, in the judgment of many, no later than eighth grade. However, it would be wrong for university and college teachers merely to excuse their own failures by a comforting reference to defects in the student's precollege education.

Today, mathematics is generally recognized as a subject of fundamental importance because of its ubiquitous role in contemporary life. To be used effectively, its methods must be understood properly—otherwise we cast ourselves in the roles of (inefficient) robots when we try to use mathematics, and we place undue strain on our naturally imperfect memories. Dr. Solow has given much thought to the question of how an understanding of a mathematical proof can be acquired. Most students today do not acquire that understanding, and Dr. Solow's plan to remedy this very unsatisfactory situation deserves a fair trial.

Peter Hilton
Distinguished Professor of Mathematics
State University of New York
Binghamton, NY

TO THE STUDENT

After finishing my undergraduate degree, I began to wonder why learning theoretical mathematics had been so difficult. As I progressed through my graduate work, I realized that mathematics possessed many of the aspects of a game: a game in which the rules had been partially concealed. Imagine trying to play chess before you know how all of the pieces move! It is no wonder that so many students have had trouble with abstract mathematics.

This book describes some of the rules by which the game of theoretical mathematics is played. It has been my experience that virtually anyone who is motivated and who has a knowledge of high school mathematics can learn these rules. Doing so will greatly reduce the time (and frustration) involved in learning abstract mathematics. I hope this book serves that purpose for you.

To play chess, you must first learn how the individual pieces move. Only after these rules have entered your subconscious can your mind turn its full attention to the more creative issues of strategy, tactics, and the like. So it appears to be with mathematics. Hard work is required in the beginning to learn the fundamental rules presented in this book. In fact, your goal should be to absorb this material so that it becomes second nature to you. Then you will find that your mind can focus on the creative aspects of mathematics. These rules are no substitute for creativity, and this book is not meant to teach creativity. However, I do believe that it can provide you with the tools needed to *express* your creativity. Equally important is the fact that these tools will enable you to understand and to appreciate the creativity of others.

You are about to learn a key part of the mathematical thought process. As you study the material and solve problems, be conscious of your own thought process. Ask questions and seek answers. Remember, the only unintelligent question is the one that goes unasked.

Cleveland, Ohio
January 1990

Daniel Solow

TO THE INSTRUCTOR

THE OBJECTIVE OF THIS BOOK

The inability to communicate proofs in an understandable manner has plagued students and teachers in all branches of mathematics. The result has been frustrated students, frustrated teachers and, oftentimes, a "watered down" course to enable the students to follow at least some of the material, or a test that protects students from the consequences of this deficiency in their mathematical understanding.

One might conjecture that most students simply cannot understand abstract mathematics, but my experience indicates otherwise. What seems to have been lacking is a proper method for explaining theoretical mathematics. In this book I have developed a method for communicating proofs: a common language that can be taught by professors and understood by students. In essence, this book categorizes, identifies, and explains (at the student's level) the various techniques that are used repeatedly in virtually all proofs.

Once the students understand the techniques, it is then possible to explain any proof as a sequence of applications of these techniques. In fact, it is advisable to do so because the process reinforces what the students have learned in the book.

Explaining a proof in terms of its component techniques is not difficult, as is illustrated in the examples of this book. Before each "condensed" proof is an analysis of the proof explaining the methodology, thought process, and techniques that are being used. Teaching proofs in this manner requires nothing more than preceding each step of the proof with an indication of which technique is about to be used, and why.

When discussing a proof in class, I actively involve the students by soliciting their help in choosing the techniques and in designing the proof. I have been pleasantly surprised by the quality of their comments and questions. It has been my experience that once the students become comfortable with the proof techniques, their minds tend to address the more important issues of mathematics, such as why a proof is done in a particular way and why the piece of mathematics is important in the first place. This book is not meant to teach creativity, but I do believe that it does describe many of the necessary underlying skills whose acquisition will free the student's mind to focus on the creative aspects. I have also found that, by using this approach, it is possible to teach subsequent mathematical material at a much more sophisticated level without losing the students.

In any event, the message is clear. I am suggesting that there are many benefits to be gained by teaching mathematical thought processes in addition to mathematical material. This book is designed to be a major step in the right direction by making abstract mathematics understandable and enjoyable to the students and by providing you with a method for communicating with them.

WHY A SECOND EDITION

After teaching the material in this book for more than 10 years to undergraduate mathematics and computer science majors and to graduate students in operations research, I have experienced how effective this systematic method for learning mathematical proofs can be. At the same time I have learned how to improve the approach significantly. This second edition incorporates those improvements as well as many of the comments I received from various colleagues via a questionnaire distributed by John Wiley. In describing the changes that follow, I assume you are familiar with the previous edition of this book.

CONTENT AND PRESENTATION

1. All "analyses of proofs" are presented in the same consistent manner—with forward statements being labeled **A, A1, A2,** and so on and backward statements being labeled **B, B1, B2,** and so on.
2. A new chapter (Chapter 8) has been included on "nested" quantifiers, that is, statements containing more than one quantifier.
3. The technique of a "proof by cases" has been included in Chapter 12.
4. Discussions have been improved on the topics that have typically caused students much difficulty, including, but not limited to:

   a. Explaining how the truth table for "A implies B" in Chapter 1 is used when the truth of the individual statements A and/or B is not immediately evident.

   b. Explaining how to overcome notational difficulties that arise when one applies a definition containing one set of symbols to a specific statement using an "overlapping" set of symbols whose meanings are different from those in the definition (see Chapter 3).

   c. Explaining that when the construction method is used (see Chapter 4), the proof consists of showing that the constructed objects satisfy the needed properties, rather than describing *how* the objects are constructed.

   d. Explaining more clearly that when the choose method is used (see Chapter 5), the chosen object—together with its properties—provides information that can be used in the forward process; while the next statement in the backward process is trying to show that, for the chosen object, the something happens.

e. Explaining more carefully how the specialization method is used (see Chapter 7).

CHANGES IN THE EXERCISES

5. The number of exercises has been approximately doubled, and the solutions to the odd numbered exercises *only* are included in the book. (Some of these exercises are more difficult but, since a large number of users of this book are first-year undergraduates, no mathematics beyond high school is required.)
6. Numerous exercises pertaining to reading condensed proofs are included, and the students are asked to explain those proofs in terms of the techniques they have learned in this book.

CHANGES IN THE TERMINOLOGY

7. The term "Outline of Proof" has been changed to "Analysis of Proof" because an "outline" suggests a shortened version of a proof rather than an expanded version, which the analysis of proof really is.
8. The term "abstraction question" has been replaced with "key question," since the latter is easier for students to relate to and to pronounce. It is also easier to talk about the "key question" and its associated "key answer."
9. The two uniqueness methods (see Chapter 12) have been given the names of the "direct" and "indirect" uniqueness methods.

10. The either/or method (see Chapter 12) applied in the forward process is referred to as a "proof by cases"; the either/or method applied in the backward process is called a "proof by elimination."

CHANGES IN THE PRODUCTION AND DISPLAY

11. Section headings are provided to delineate the various topics within each chapter, and a final summary section is included in each chapter.
12. All of the material is professionally typeset.

Although these changes seem to make it even easier for students to understand the meaning of a mathematical proof, I have still found no substitute for actively teaching the material in class instead of having the students read the material on their own. This active interaction has proved eminently beneficial to both student and teacher, in my case.

Cleveland, Ohio
January 1990

Daniel Solow

ACKNOWLEDGMENTS

For helping to get this work known in the mathematics community, my deepest gratitude goes to Peter Hilton, an outstanding mathematician and educator. I also thank Paul Halmos, whose timely recognition and support greatly facilitated the dissemination of the knowledge of the existence of this book and teaching method. I am also grateful for discussions with Gail Young and George Polya.

Regarding the preparation of the manuscript for the first edition, no single person had more constructive comments than Tom Butts. He not only contributed to the mathematical content but also corrected many of the grammatical and stylistic mistakes in a preliminary version. I suppose that I should thank his mother for being an English teacher. I also acknowledge Charles Wells for reading and commenting on the first handwritten draft and for encouraging me to pursue the project further. Many other people made substantive suggestions, including Alan Schoenfeld, Samuel Goldberg, and Ellen Stenson.

Of all the people who were involved in this project, none deserve more credit than my students. It is because of their voluntary efforts that the first edition of this book was prepared in such a short time. Thanks especially to John Democko for acting in the capacity of senior editor while concurrently trying to complete his Ph.D. program. Also, I appreciate the help that I received from Michael Dreiling and Robert Wenig in data basing the text and in preparing the exercises. Michael worked on this project almost as long as I did. A special word of thanks goes to Greg Madey for coordinating the second rewriting of the document, for adding useful comments and, in general, for keeping me very organized. His responsibilities were subsequently assumed by Robin Symes.

In addition, I am grateful to Ravi Kumar for the long hours he spent on the computer preparing the final version of the manuscript, to Betty Tracy and Martha Bognar for their professional and flawless typing assistance, and to Virginia Benade for her technical editing. I am also indebted to my class of 1981 for preparing the solutions to the homeworks.

I thank the following professors for refereeing the original manuscript and for recommending its publication: Alan Tucker, David Singer, Howard Anton, and Ivan Niven.

Most of the credit for the improvements in this second edition go to the students who have taught me so much about learning mathematics. I am also grateful for the many comments and suggestions I have received from colleagues over the years, and more recently on a questionnaire circulated by John Wiley.

On the technical end, I thank Dawnn Strasser for her most professional job of typing the preliminary version of the manuscript. I also am grateful to the Weatherhead School of Management at Case Western Reserve University for the use of their excellent word processing facilities.

Finally, I am grateful to my wife, Audrey, for her help in proofreading and for her patience during yet another of my projects.

D. S.

CONTENTS

LIST OF TABLES

LIST OF DEFINITIONS

ONE

THE TRUTH OF IT ALL

The objective of mathematicians is to discover and to communicate certain truths. Mathematics is the language of mathematicians, and a proof is a method of communicating a mathematical truth to another person who also "speaks" the language. A remarkable property of the language of mathematics is its precision. Properly presented, a proof will contain no ambiguity: there will be no doubt as to its correctness. Unfortunately, many proofs that appear in textbooks and journal articles are not presented properly; more appropriately stated, the proofs are presented properly for someone who already knows the language of mathematics. Thus, to understand and/or present a proof, you must learn a new language, a new method of thought. This book explains much of the basic "grammar" you will need but, as in learning any new language, a lot of practice on your part will be needed to become fluent.

THE OBJECTIVES OF THIS BOOK

The approach of this book is to categorize and to explain the various techniques that are used in *all* proofs, independent of the subject matter. One objective is to teach you how to read and understand a written proof by identifying the techniques that have been used. Learning to do so will enable you to study almost any mathematical subject on your own, a desirable goal in itself.

A second objective of this book is to teach you to develop and to communicate your own proofs of known mathematical truths. Doing so requires that you use a certain amount of creativity, intuition, and experience. Just as there are many ways to express the same idea in any language, so are there different proofs for the same mathematical fact. The proof techniques presented here are designed to get you started and to guide you through a proof. Consequently, this book describes not only *how* the proof techniques work, but also, *when* each technique is likely to be used, and *why*. It is often the case that a correct technique can be chosen based on the very form of the problem under consideration. Therefore, when attempting to create your own proof, *learn to select a proof technique consciously* before wasting hours trying to figure out what to do. The more aware you are of your thought process, the better it is.

The ultimate objective, however, is to use your newly acquired skills and language to discover and communicate previously unknown mathematical proofs. While the goal is an admirable one, it is quite difficult to attain. The first step in this direction is to reach the level of being able to read proofs and to develop your own proofs of already-known facts. This alone will give you a much deeper and richer understanding of the mathematical universe around you.

The basic material on proof techniques is presented in the first twelve chapters. The thirteenth chapter is a complete summary and it is followed by two appendices that illustrate the various techniques with several examples.

The book can be read by anyone with a good knowledge of high school mathematics. Advanced students who have seen proofs before can read the first two chapters, skip to the summary chapter, and subsequently read the two appendices to see how all the techniques fit together. The remainder of this chapter explains the types of relationships to which proofs can be applied.

WHAT IS A PROOF?

In mathematics, a **statement** is an expression that is either true or false. Some examples of statements are:

1. Two different lines in a plane are either parallel or else they intersect in exactly one point.
2. $1 = 0$.
3. $3x = 5$ and $y = 1$.
4. x is not > 0.
5. There is an angle t such that $\cos(t) = t$.

Observe that statement (1) is always true, (2) is always false, and statements (3) and (4) can be either true or false depending on the value of a variable.

It is perhaps not as obvious that statement (5) is always true. It therefore becomes necessary to have some method for "proving" that such statements are true. In other words, a **mathematical proof** is a convincing argument expressed in the language of mathematics. In this and other books, proofs are often given for what seem to be obviously true statements. One reason for doing this is to provide examples that are easy to follow so that you can develop techniques for proving more difficult statements. Another reason for doing this is that some apparently "obvious" statements are, in fact, false. You will know that a statement is true only when you have *proved* it to be true.

A proof should contain enough mathematical detail so as to be convincing to the person(s) to whom the proof is addressed. A proof of statement (5) above aimed at convincing a mathematics professor might consist of nothing more than Figure 1.1; whereas a proof directed toward a high school student would probably require more detail, perhaps even the definition of cosine. Your proofs should contain enough detail to be convincing to someone else at your own mathematical level (e.g., a classmate). It is the lack of sufficient detail that can often make a proof difficult to

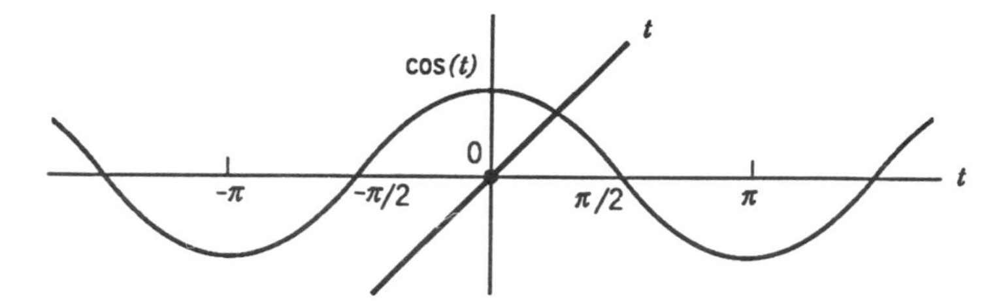

Figure 1.1 A Proof That There is an Angle t Such That $\cos(t) = t$

read and understand. One objective of this book is to teach you to decipher these "condensed" proofs that are likely to appear in textbooks and other mathematical literature.

Given two statements A and B, each of which may be either true or false, a fundamental problem of interest in mathematics is to show that

If A is true then B is true.

One reason for wanting to do so arises when B is a statement you would *like* to be true but whose truth is not easy to verify. In contrast, suppose that A is statement whose truth is relatively easy to verify. If you have proved that "if A is true then B is true," and if you can verify that A is in fact true, then you will know that B is true.

A proof is a formal method for convincing yourself (and others) that "if A is true then B is true." In order to do a proof, you must know exactly what it means to show that "if A is true then B is true." The statement A is often called the **hypothesis** and B the **conclusion**. For brevity, the statement "if A is true then B is true" is shortened to "if A then B" or simply "A implies B." Mathematicians are often very lazy when it comes to writing, so they have developed a symbolic "shorthand." For instance, a mathematician would write "$A \Rightarrow B$" instead of "A implies B." For the most part, textbooks do not use the symbolic notation but teachers often do, and eventually you may find it useful too. Therefore this book will include the appropriate symbols but will not use them in proofs. A complete list of the symbols can be found in the glossary at the end of this book.

It seems reasonable that the conditions under which "A implies B" are true will depend on whether A and B themselves are true. Consequently, there are four possible cases to consider:

1. A is true and B is true.
2. A is true and B is false.
3. A is false and B is true.
4. A is false and B is false.

Suppose, for example, that your friend made the statement: "If it rains then Mary brings her umbrella." Here, the statement A is "it rains" and B is "Mary brings her umbrella." To determine when the statement "A implies B" is false, ask yourself in which of the four cases above would you be willing to call your friend a liar. In the first case (i.e., when it *does* rain and Mary *does* bring her umbrella) your friend has told the truth. In the second case, it has rained, and yet Mary did *not* bring her umbrella, as your friend said she would. Here your friend has not told the truth. In cases (3) and (4), it does not rain. You would not really want to call your friend a liar in the case of no rain because, your friend only said that something would happen if it *did* rain. Thus, the statement "A implies B" is true in each of the four cases except the second one, as summarized in Table 1.

Table 1 is an example of a **truth table**, which is a method for determining when a complex statement (in this case, "A implies B") is true by examining all possible truth values of the individual statements (in this case, A and B). Other examples of truth tables will appear in Chapter 3.

Table 1. The Truth of "A Implies B"

A	B	A Implies B
True	True	True
True	False	False
False	True	True
False	False	True

According to Table 1, when trying to show that "A implies B" is true, you might attempt to determine the truth of A and B individually, and then use the appropriate row of Table 1 to determine the truth of "A implies B." For example, to determine the truth of the statement:

If $1 < 2$ then $4 < 3$

you can easily see that the hypothesis A (i.e., $1 < 2$) is true and the conclusion B (i.e., $4 < 3$) is false. Thus, using the second row of Table 1 (corresponding to A being true and B being false) you can conclude that, in this case, the statement "A implies B" is false. Similarly, the statement:

If $2 < 1$ then $3 < 4$

is true by the third row of Table 1 because A (i.e., $2 < 1$) is false and B (i.e., $3 < 4$) is true.

Now suppose that you want to prove that the following statement is true:

If $x > 2$ then $x^2 > 4$.

The difficulty with using Table 1 for this example is that you cannot determine whether A (i.e., $x > 2$) and B (i.e., $x^2 > 4$) are true or false — this is because the truth of A and B depend on the variable x whose value is not known. Nonetheless, you can use Table 1 by reasoning as follows:

> Although I do not know the truth of A, I do know that it must be either true or false. Let me assume, for the moment, that A is false (subsequently, I will consider what happens when A is true). When A is false, either the third or the fourth row of Table 1 is applicable and, in either case, the statement "A implies B" will be true — thus I would be done. Therefore, *I need only consider the case in which A is true.*
>
> In the case when A is true, either the first or the second row of Table 1 is applicable. However, since I want to prove that the statement "A implies B" is true, I need to be sure that the first row of the truth table is applicable, and *this can be done by establishing that B is true.*

As a result of the above reasoning, when trying to prove that "A implies B" is true, *you can assume that A is true; your job is to conclude that B is true.*

Note that a proof of the statement "A implies B" is *not* an attempt to verify whether A and B themselves are true, but rather, to show that B is a logical result of having *assumed* that A is true. Your ability to show that B is true will depend on the fact that you have assumed A to be true; ultimately, you will have to discover the relationship between A and B. Doing so requires a certain amount of creativity. The proof techniques presented here are designed to get you started and guide you along the path.

SUMMARY

Hereafter, A and B will be statements that are either true or false. The problem of interest will be that of showing "A implies B." To do so,

1. Assume that A is true.
2. Use this assumption to reach the conclusion that B is true.

EXERCISES

1.1. Which of the following are mathematical statements? (Recall that a mathematical statement must be either true or false.)

a. $ax^2 + bx + c = 0$.

b. $\dfrac{(-b + \sqrt{b^2 - 4ac})}{(2a)}$.

c. Triangle XYZ is similar to triangle RST.

d. $3 + n + n^2$.

e. $\sin(\frac{\pi}{2}) < \sin(\frac{\pi}{4})$.

f. For every angle t, $\sin^2(t) + \cos^2(t) = 1$.

1.2. Which of the following mathematical statements are true?

a. The square root of any integer is a nonnegative real number.

b. There is an angle t such that $\sin(t) = \cos(t)$.

c. $x < 1$.

d. If $x < 1$ then $x^2 < 1$.

1.3. For each of the following problems, identify the hypothesis and the conclusion.

a. If the right triangle XYZ with sides of lengths x and y, and hypotenuse of length z has an area of $\frac{z^2}{4}$, then the triangle XYZ is isosceles.

b. n is an even integer $\Rightarrow n^2$ is an even integer.

c. If a, b, c, d, e, and f are real numbers with the property that $ad - bc \neq 0$, then the two linear equations $ax + by = e$ and $cx + dy = f$ can be solved for x and y.

d. The sum of the first n positive integers is $\frac{n(n+1)}{2}$.

e. r is real and satisfies $r^2 = 2$ implies r is irrational.

f. If p and q are positive real numbers with $\sqrt{pq} \neq \frac{(p+q)}{2}$ then $p \neq q$.

g. When x is a real number, the minimum value of

$x(x-1)$ is at least $-\frac{1}{4}$.

1.4. "If I do not get a promotion, my wife will be unhappy," says Jack. Later, you come to know that Jack has been promoted, but you find that his wife is unhappy. Was Jack's statement true or false? Explain.

1.5. Using Table 1 on page 6, determine the conditions on A and B under which the following statements are true or false and give your reason.

a. If $2 > 7$ then $1 > 3$.

b. If $2 < 7$ then $1 < 3$.

c. If $x = 3$ then $1 < 2$.

d. If $x = 3$ then $1 > 2$.

1.6. Suppose someone says to you that the following statement is true: "If Mr. Smith wins the election, then you are your own child." Using Table 1 on page 6, did Mr. Smith win the election? Why or why not? Explain.

1.7. If you are trying to prove that "A implies B" is true and you know that B is false, do you want to show that A is true or false? Explain.

1.8. By considering what happens when A is true and when A is false, it was decided that when trying to prove the statement "A implies B" is true, you can assume that A is true and your goal is to show that B is true. Derive another approach to proving "A implies B" is true by considering what happens when B is true and when B is false. What should you assume is true and what should you try to conclude is true?

1.9. Using Table 1 on page 6, prepare a truth table for each of the following statements.

a. A implies (B implies C).

b. (A implies B) implies C.

1.10. a. Suppose you want to show that "A implies B" is *false*. According to Table 1 on page 6, how should you do this? What should you try to show about the truth of A and B?

b. Apply your answer of part (a) to show that the statement "If x is a real number that satisfies $-3x^2 + 2x + 8 = 0$ then $x > 0$" is false.

TWO

THE FORWARD-BACKWARD METHOD

The purpose of this chapter is to describe one of the fundamental proof techniques: the **forward–backward method**. Special emphasis is given to the material of this chapter because all other proof techniques will rely on this method.

The first step in proving that "If A then B" is true requires recognizing the statements A and B. In general, everything after the word "if" and before the word "then" comprises statement A; everything after the word "then" constitutes statement B. Alternatively, everything that you are assuming to be true (i.e., the hypothesis) is A; everything that you are trying to prove (i.e., the conclusion) is B. Consider the following example.

EXAMPLE 1.

Proposition. If the right triangle XYZ with sides of lengths x and y, and hypotenuse of length z, has an area of $\frac{z^2}{4}$, then the triangle XYZ is isosceles (see Figure 2.1).

Analysis of Proof. In this example one has the statements:

A: The right triangle XYZ with sides of lengths x and y, and hypotenuse of length z has an area of $\frac{z^2}{4}$.

B: The triangle XYZ is isosceles.

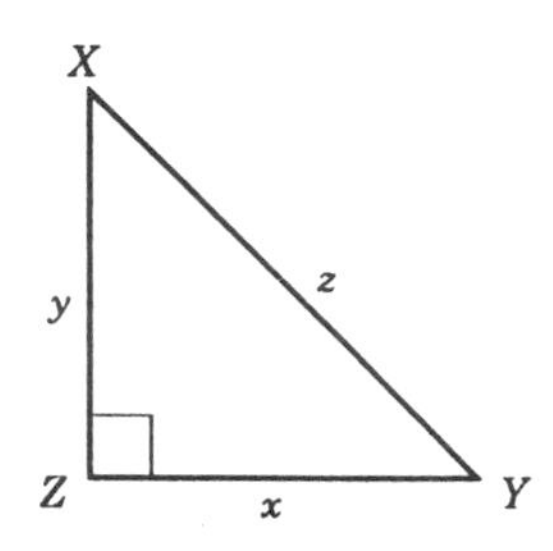

Figure 2.1 The Right Triangle XYZ

Recall from the discussion at the end of Chapter 1 that, when proving "A implies B," you can assume that A is true and you must somehow use this information to reach the conclusion that B is true. In attempting to figure out just how to reach the conclusion that B is true, you will be going through a **backward process**; when you make specific use of the information contained in A, you will be going through a **forward process**. Both of these processes will be described in detail now.

THE BACKWARD PROCESS

In the backward process you begin by asking "How or when can I conclude that the statement B is true?" The very manner in which you phrase this question is critical since you must eventually be able to answer it. The question should be posed in an *abstract* way. For Example 1, the correct (abstract) question is: "How can I show that *a* triangle is isosceles?" While it is true that you want to show that the particular triangle XYZ is isosceles, by asking the abstract question, you call on your general knowledge of triangles, clearing away irrelevant details (such as the fact that the triangle is called XYZ instead of ABC), thus allowing you to focus on those aspects of the problem that really seem to matter. The question obtained from statement B in such problems will be called the **key question**. A properly posed key question should contain no symbols or other notation (except for numbers) from the specific problem under consideration. The key to many proofs is formulating a correct key question.

Once you have posed the key question, the next step in the backward process is to answer it. Returning to Example 1, how can you show that a triangle is isosceles? Certainly one way is to show that two of its sides have equal length. Referring to Figure 2.1, you should show that $x = y$. Observe that answering the key question is a two-phase process. First you give an abstract answer that contains no symbols from the specific problem: to show that *a* triangle is isosceles, show that two of *its* sides have equal length. Next, you apply this answer to the specific situation: to show that two of its sides have equal length means to show that $x = y$ (not that $x = z$ or $y = z$). The process of asking the key question, answering it abstractly, and then applying that answer to the specific situation constitutes one step of the backward process.

The backward process has given you a new statement, $B1$, with the property that *if* you could show that $B1$ is true, then B would be true. For the example above, the new statement is

B1: $x = y$.

If you can show that $x = y$, then the triangle XYZ is isosceles. Once you have the statement $B1$, all of your efforts must now be directed toward reaching the conclusion that $B1$ is true, for then it will follow that B is true. How can you show that $B1$ is true? Eventually you will have to make use of the assumption that A is true. When solving this problem, you would most likely do so now but, for the moment, let us continue by working backward from the new statement $B1$. This will illustrate some of the difficulties that can arise in the backward process. Can you pose the new key question for $B1$?

Since x and y are the lengths of two sides of a triangle, a reasonable key question is "How can I show that the lengths of two sides of a triangle are equal?" A second perfectly reasonable key question is "How can I show that two real numbers are equal?" After all, x and y are also real numbers. One of the difficulties that can arise in the backward process is the possibility of more than one key question. Choosing the correct one is more of an art than a science. In fortunate circumstances, there will be only one obvious key question; in other cases you may have to proceed by trial and error. This is where your intuition, insight, creativity, experience, diagrams, and graphs can play an important role. One general guideline is to let the information in A (which you are assuming to be true) help you to choose the question, as will be done in this case. Regardless of which question you finally settle on, the next step will be to answer it, first in the abstract and then in the specific situation. Can you do this for the two key questions posed above? For the first one, you might show that two sides of a triangle have equal length by showing that the angles opposite them are equal. For the triangle XYZ of Figure 2.1, this would mean you have to show that angle X equals angle Y. A cursory examination of the contents of statement A does not seem to provide much information concerning the angles of triangle XYZ. For this reason, the other key question will be chosen.

Now you are faced with the question "How can I show that two real numbers (namely, x and y) are equal?" One answer to this question is to show that the difference of the two numbers is 0. Applying this answer to the specific statement $B1$ means you would have to show that $(x - y) = 0$. Unfortunately, there is another perfectly acceptable answer: show that the first number is less than or equal to the second number and also that the second number is less than or equal to the first number. Applying this answer to the specific statement $B1$, you would have to show that $x \leq y$ and $y \leq x$. Thus, a second difficulty can arise in the backward process: even if you choose the correct key question, there may be more than one answer to it. Moreover, you might choose an answer that will not permit you to complete the proof. For instance, associated with the key question "How can I show that a triangle is isosceles?" is the answer "Show that the triangle is equilateral." Of course it will be impossible to show that triangle XYZ of Example 1 is equilateral since one of its angles is 90 degrees.

Returning to the key question associated with $B1$, "How can I show that two real numbers (namely, x and y) are equal?" suppose, for the sake of argument, that you choose the answer of showing that their difference is 0. Once again, the backward process has given you a new statement, $B2$, with the property that if you could show that $B2$ is true, then in fact $B1$ would be true, and hence so would B. Specifically, the new statement is:

B2: $x - y = 0$.

Now all of your efforts must be directed toward reaching the conclusion that $B2$ is true. You must ultimately make use of the information in A but, for the moment, let us continue once more with the backward process applied to $B2$.

One key question associated with B2 is: "How can I show that the difference of two real numbers is 0?" After some reflection, it may seem that there is no reasonable answer to this question. Yet another problem can arise in the backward process: the key question might have no apparent answer! Do not despair – all is not lost. Remember that when proving "A implies B," you are allowed to assume that A is true. It is now time to make use of this fact.

THE FORWARD PROCESS

The forward process involves starting with the statement A, which you assume to be true, and deriving from it some other statement, $A1$, which you know to be true as a result of A being true. It should be emphasized that the statements derived from A are not haphazard. Rather, they are directed toward linking up with the last statement obtained in the backward process. Let us return to Example 1, keeping in mind the fact that the last statement in the backward process is $B2$: $x - y = 0$.

For Example 1 above, the statement A is "The right triangle XYZ with sides of length x and y, and hypotenuse of length z, has an area of $\frac{z^2}{4}$." One fact that you know (or should know) as a result of A being true is that $\frac{xy}{2} = \frac{z^2}{4}$, because the area of a triangle is one-half the base times the height, in this case $\frac{xy}{2}$. So you have obtained the new statement:

$$\textbf{A1:} \quad \frac{xy}{2} = \frac{z^2}{4}.$$

Another useful statement follows from A by the Pythagorean theorem because XYZ is a right triangle, so you also have:

$$\textbf{A2:} \ (x^2 + y^2) = z^2.$$

The forward process can also combine and use the new statements to produce more true statements. For instance, it is possible to combine A1 and A2 by replacing z^2 in $A1$ with $(x^2 + y^2)$ from $A2$ obtaining the statement:

$$\textbf{A3:} \quad \frac{xy}{2} = \frac{(x^2 + y^2)}{4}.$$

One of the problems with the forward process is that it is also possible to generate some useless statements, for instance, "angle X is less than 90 degrees." While there are no specific guidelines for producing new statements, keep in mind the fact that the forward process is directed toward obtaining the statement $B2$: $x - y = 0$, which was the last one derived in the backward process. The fact that $B2$ does not contain the quantity z is the reason that z^2 was eliminated from $A1$ and $A2$ to produce $A3$.

Continuing with the forward process, you should attempt to rewrite $A3$ so as to make it look more like $B2$. For instance, you can multiply both sides of $A3$ by 4 and subtract $2xy$ from both sides to obtain:

$$\textbf{A4:} \quad (x^2 - 2xy + y^2) = 0.$$

Factoring $A4$ yields:

$$\textbf{A5:} \quad (x - y)^2 = 0.$$

One of the most common steps of the forward process is to rewrite statements in different forms, as was done in obtaining $A4$ and $A5$. For Example 1, the final step in the forward process (and in the entire proof) is to realize from $A5$ that if the square of a number (namely, $x - y$) is 0, then the number itself is 0, thus obtaining precisely the statement $B2$: $x - y = 0$. The proof is now complete since you started with the assumption that A is true and used it to derive the conclusion that $B2$, and hence B, is true. The steps and reasons are summarized in Table 2.

It is interesting to note that the forward process ultimately produced the elusive answer to the key question associated with $B2$: "How can I show that the difference of two real numbers is 0?"– which, in this case, is to show that the square of the difference is 0 (see $A5$ in Table 2).

Table 2. Proof of Example 1

	Statement	Reason
A :	Area of XYZ is $\frac{z^2}{4}$	Given
A1:	$\frac{xy}{2} = \frac{z^2}{4}$	Area = (base)(height)/2
A2:	$x^2 + y^2 = z^2$	Pythagorean theorem
A3:	$\frac{xy}{2} = (x^2 + y^2)/4$	Substitute $A2$ into$A1$
A4:	$x^2 - 2xy + y^2 = 0$	From $A3$ by algebra
A5:	$(x - y)^2 = 0$	Factoring $A4$
B2:	$(x - y) = 0$	From $A5$ by algebra
B1:	$x = y$	Add y to both sides of $B2$
B :	XYZ is isosceles	Since $B1$ is true

Finally, you should realize that in general it will not be practical to write down the entire thought process that goes into a proof, for this would require too much time, effort, and space. Rather, a highly condensed version is usually presented and often makes little or no reference to the backward process. For the problem above the condensed proof might go something like this.

Proof of Example 1. From the hypothesis and the formula for the area of a right triangle, the area of $XYZ = \frac{xy}{2} = \frac{z^2}{4}$. By the

Pythagorean theorem, $(x^2+y^2) = z^2$, and on substituting (x^2+y^2) for z^2 and performing some algebraic manipulations one obtains $(x-y)^2 = 0$. Hence $x = y$ and the triangle XYZ is isosceles. ∎

(The ∎ or some equivalent symbol is usually used to indicate the end of the proof. Sometimes the letters Q.E.D. are used as they stand for the Latin words *quod erat demonstrandum*, meaning "which was to be demonstrated.")

Sometimes the condensed proof will be partly backward and partly forward. For example:

Proof of Example 1. The statement will be proved by establishing that $x = y$, which in turn is done by showing that $(x-y)^2 = (x^2 - 2xy + y^2) = 0$. But the area of the triangle is $\frac{xy}{2} = \frac{z^2}{4}$, so that $2xy = z^2$. By the Pythagorean theorem, $z^2 = (x^2 + y^2)$ and hence $(x^2 + y^2) = 2xy$, or $(x^2 - 2xy + y^2) = 0$. ∎

The proof can also be written entirely from the backward process. Although slightly unnatural, this version is worth seeing.

Proof of Example 1. To reach the conclusion, it will be shown that $x = y$ by verifying that $(x-y)^2 = (x^2 - 2xy + y^2) = 0$, or equivalently, that $(x^2 + y^2) = 2xy$. This can be established by showing that $2xy = z^2$, for the Pythagorean theorem states that $(x^2 + y^2) = z^2$. In order to see that $2xy = z^2$, or equivalently, that $\frac{xy}{2} = \frac{z^2}{4}$, note that $\frac{xy}{2}$ is the area of the triangle and it is equal to $\frac{z^2}{4}$ by hypothesis, thus completing the proof. ∎

Proofs found in research articles are very condensed, giving little more than a hint of how to do the proof. For example:

Proof of Example 1. The hypothesis together with the Pythagorean theorem yield $(x^2 + y^2) = 2xy$; hence $(x-y)^2 = 0$. Thus the triangle is isosceles as required. ∎

Note that the word "hence" effectively conceals the reason that $(x-y) = 0$. Was it algebraic manipulation (as we know it was) or something else? These highly condensed versions are typically given in mathematics books. From these examples, you can see that there are several reasons why reading condensed proofs can be challenging:

1. The steps of the proof may not be presented in the same order in which they were performed when the proof was done (see, for example, the first condensed proof of Example 1 given above).
2. The names of the techniques are often omitted (for instance, in the condensed proofs given above, no mention was made of the forward and backward processes or the key question).
3. Several steps of the proof are often combined into a single statement with little or no explanation (as was done in the last condensed proof of Example 1 given above).

You should strive toward the ability to read and to dissect a condensed proof. To do so, you will have to figure out which proof techniques are being used (since the forward-backward method is not the only one available). Then, from what is written, you will have to discover the thought process that went into the proof, and finally, be able to verify all the steps involved. The more condensed the proof, the harder this process will be. When an author writes "It is easy to see that ...," or "Clearly, ...," you can assume it will take you quite some time to fill in the missing details.

Some examples of how to read condensed proofs appear in the two appendices. In this book your life will be made much easier because preceding each condensed proof will be an analysis that describes the techniques, methodology, and reasoning involved in doing the proof. Out of necessity, these "Analysis-of-Proof" discussions will be more succinct than the one given in Example 1.

SUMMARY

A summary of the forward–backward method for proving that "*A*

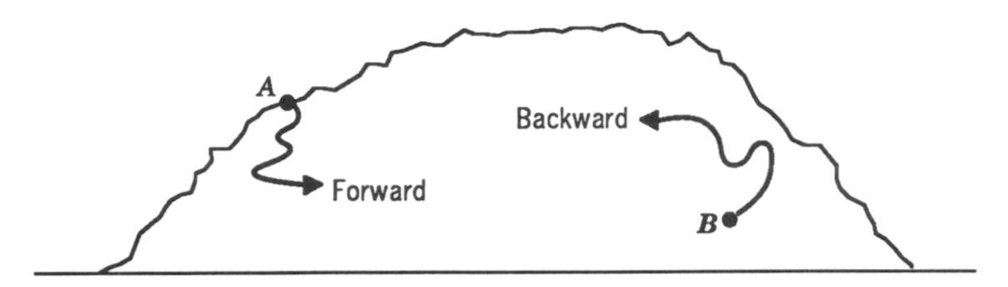

Figure 2.2 Finding a Needle in a Haystack

implies B" is in order. Begin with the statement B which you are trying to conclude is true. Through the backward process of asking and answering the key question, derive a new statement, $B1$, with the property that if $B1$ is true, then so is B. All efforts are now directed toward establishing that $B1$ is true. To that end, apply the backward process to $B1$ obtaining a new statement, $B2$, with the property that if $B2$ is true, then so is $B1$ (and hence B). Remember that the backward process is motivated by the fact that A is assumed to be true. Continue in this manner until either you obtain the statement A (in which case the proof is finished) or until you can no longer pose and/or answer the key question fruitfully. In the latter case, it is time to start the forward process in which you derive a sequence of statements from A that are necessarily true as a result of A being assumed true. Remember that the goal of the forward process is to obtain precisely the last statement you had in the backward process, at which time you will have successfully completed the proof.

These two processes can easily be remembered by thinking of the statement B as a needle in a haystack. When you work forward from the assumption that A is true, you start somewhere on the outside of the haystack and try to find the needle. In the backward process, you start at the needle and try to work your way out of the haystack toward the statement A (see Figure 2.2).

Another way of remembering the forward–backward method is to think of a maze in which A is the starting point and B is the ending point (see Figure 2.3). You may have to alternate several times between the forward and backward processes because there

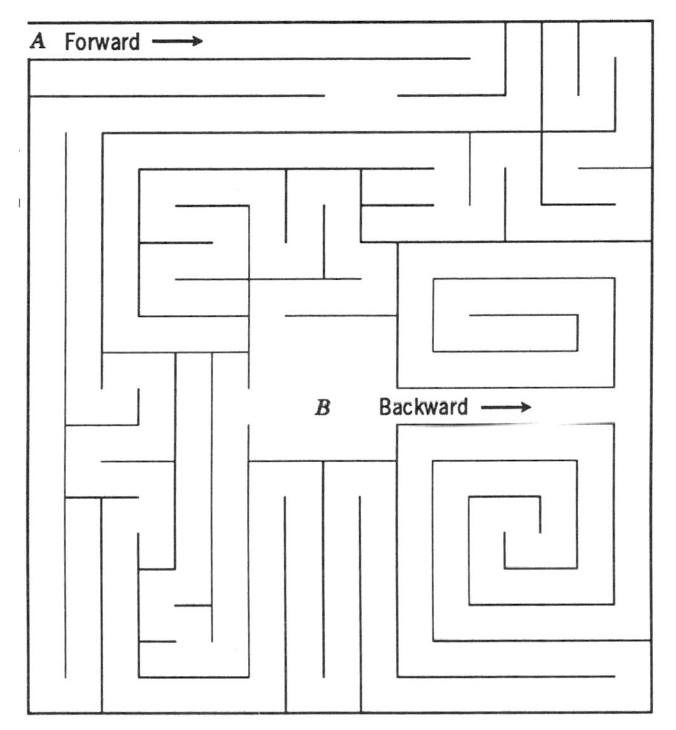

Figure 2.3 The Maze

are likely to be several false starts and blind alleys.

As a general rule, the forward–backward method is probably the first technique to try on a problem unless you have reason to use a different approach based on the form of B, as will be described shortly. In any case, you will gain much insight into the relationship between A and B.

EXERCISES

Note: All proofs should contain an analysis of proof as well as a condensed version.

2.1. Explain the difference between the forward and backward processes. Describe how each one works and what can go wrong. How are the two processes related to each other?

2.2. When formulating the key question, should you look at the last statement in the forward process or in the backward process? When answering the key question, should you be guided by the last statement in the forward process or in the backward process?

2.3. Consider the problem of proving that "If x is a real number, then the maximum value of $-x^2 + 2x + 1$ is greater than or equal to 2." Which of the following key questions is incorrect? Explain.

a. How can I show that the maximum value of a parabola is greater than or equal to a number?

b. How can I show that a number is less than or equal to the maximum value of a polynomial?

c. How can I show that the maximum value of the function $-x^2+2x+1$ is greater than or equal to a number?

d. How can I show that a number is less than or equal to the maximum of a quadratic function?

2.4. For the key question: "How can I show that two triangles are congruent?" which of the following answers is incorrect? Explain.

a. Show that two angles and the side connecting those two angles of one triangle are equal to the corresponding angles and side of the other triangle.

b. Show that the lengths of all three sides of one triangle are equal to the lengths of the corresponding sides of the other triangle.

c. Show that all three angles of one triangle are equal to the corresponding angles of the other triangle.

d. Show that the lengths of two sides and the angle between those sides in one triangle are equal to the corresponding sides and angle in the other triangle.

2.5. Consider the problem of showing that "If

$R = \{$ real numbers $x : x^2 - x \leq 0\}$,
$S = \{$ real numbers $x : -(x-1)(x-3) \geq 0\}$, and
$T = \{$ real numbers $x : x \geq 1\}$,

then R intersect S is a subset of T." Which of the following key questions is correct, and why? Explain what is wrong with the other choices.

a. How can I show that a set is a subset of another set?

b. How can I show that R intersect S is a subset of T?

c. How can I show that every point in R intersect S is greater than or equal to 1?

d. How can I show that the intersection of two sets has a point in common with another set?

2.6. Suppose you are trying to prove that: "If l_1 and l_2 are tangent lines to the endpoints e_1 and e_2 of a diameter d of a circle C, then l_1 and l_2 are parallel." What is wrong with the key question: "How can I show that two lines are tangent to the endpoints of a diameter of a circle?"

2.7. For each of the following problems, list as many key questions as you can (at least two). Be sure your questions contain no symbols or notation from the specific problem.

a. If l_1 and l_2 are the tangent lines to a circle C at the two endpoints e_1 and e_2 of a diameter d, respectively, then l_1 and l_2 are parallel.

b. If f and g are continuous functions then the function $f+g$ is continuous. (Note: Continuity is a property of a function.)

c. If n is an even integer then n^2 is an even integer.

d. If n is a given integer satisfying $-3n^2 + 2n + 8 = 0$, then $2n^2 - 3n = -2$.

2.8. For each of the following problems, list as many key questions as you can (at least two). Be sure your questions contain no symbols or notation from the specific problem.

a. If a and b are real numbers then $|a+b| \leq |a|+|b|$.

b. If $y = m_1x + b_1$ and $y = m_2x + b_2$ are the equations of two lines whose slopes are different, then the two lines intersect.

c. If a and b are nonnegative real numbers then $(a+b)/2 \geq \sqrt{ab}$.

d. If R and S are the sets described in Exercise 2.5, then R intersect S is not the empty set.

2.9. For each of the following key questions, list as many answers as you can (at least three).

a. How can I show that two real numbers are equal?

b. How can I show that two triangles are congruent?

c. How can I show that two different lines are parallel?

2.10. For each of the following key questions, list as many answers as you can (at least three).

a. How can I show that a quadrilateral is a rectangle?

b. How can I show that two lines are perpendicular?

c. How can I show that two sets are equal?

2.11. For each of the following problems, (1) pose a key question, (2) answer it abstractly, and (3) apply your answer to the specific problem.

a. If a, b, and c are real numbers for which $a > 0$, $b < 0$, and $b^2 - 4ac = 0$, then the solution to the equation $ax^2 + bx + c = 0$ is positive.

b. In the following diagram, if SU is a perpendicular bisector of RT, and $\overline{RS} = 2\overline{RU}$, then triangle RST is equilateral.

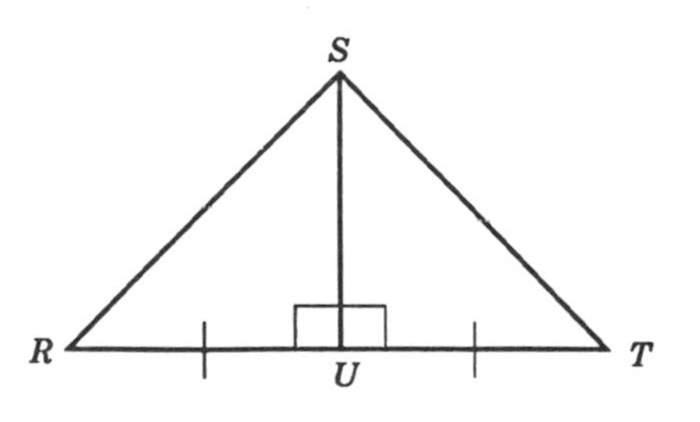

2.12. For the triangles RSU and SUT in the figure for Exercise 2.11(b), suppose you have asked the key question "How can I show that two triangles are congruent?" What is wrong with the answer "Show that $\overline{RS} = \overline{TS}$, $\angle RUS = \angle SUT$ and $\overline{US} = \overline{US}$.

2.13. For each of the following hypotheses, list as many statements as you can (at least two) that are a result of applying the forward process one step from the hypothesis.

a. The real number x satisfies $x^2 - 3x + 2 < 0$.

b. The sine of angle X in triangle XYZ of Figure 2.1 is $1/\sqrt{2}$.

c. The circle C consists of all values for x and y that satisfy $(x-3)^2 + (y-2)^2 = 25$.

2.14. For each of the following hypotheses, list as many statements as you can (at least two) that are a result of applying the forward process one step from the hypothesis.

a. The triangle UVW is equilateral.

b. The integer $3n$ is even.

c. The function $x^2 + 2x$ intersects the line $y = 2x + 1$.

2.15. Consider the problem of proving that "If x and y are real numbers such that $x^2 + 6y^2 = 25$ and $y^2 + x = 3$, then $|y| = 2$." In working forward from the hypothesis, which of the following is not valid? Explain.

a. $y^2 = 3 - x$.

b. $y^2 = \frac{25}{6} - (\frac{x}{\sqrt{6}})^2$.

c. $(3 - y^2)^2 + 6y^2 - 25 = 0$.

d. $(x + 5) = \frac{-6y^2}{(x-5)}$.

2.16. Suppose you are trying to prove that "If R is a subset of S and S is a subset of T then R is a subset of T." What is wrong with the following statement in the forward process: "Since R is a subset of T, it follows that every element of R is also an element of T."

2.17. Consider the problem of proving that "If x and y are non-negative real numbers that satisfy $x + y = 0$, then $x = 0$ and $y = 0$."

a. For the following condensed proof, write an analysis of the proof indicating the forward and backward steps, and the key questions and answers.

Proof. First it will be shown that $x \leq 0$, for then, since $x \geq 0$ by the hypothesis, it must be that $x = 0$. To see that $x \leq 0$, by the hypothesis, $x + y = 0$, so $x = -y$. Also, since $y \geq 0$, it follows that $-y \leq 0$ and hence $x = -y \leq 0$. Finally, to see that $y = 0$, since $x = 0$ and $x + y = 0$, one has $0 = x + y = 0 + y = y$. ∎

b. Rewrite the condensed proof of part (a) entirely from the backward process.

2.18. For the problem of proving that "If n is an integer greater than 2, a and b are the lengths of the legs of a right triangle, and c is the length of the hypotenuse, then $c^n > a^n + b^n$," provide justification for each sentence in the following condensed proof.

Proof. We have that $c^n = c^2c^{n-2} = (a^2 + b^2)c^{n-2}$. Observing that $c^{n-2} > a^{n-2}$ and $c^{n-2} > b^{n-2}$, it follows that $c^n > a^2(a^{n-2}) + b^2(b^{n-2})$. Consequently, $c^n > a^n + b^n$. ∎

2.19. Consider an alphabet consisting of the two letters s and t, together with the following rules for creating new "words" from old ones. (The rules can be applied in any order.)

(1) Double the current word (e.g., *sts* could become *stssts*).

(2) Erase tt from the current word (e.g., *stts* could become *ss*).

(3) Replace *sss* in the current word by t (e.g., *stsss* could become *stt*).

(4) Add the letter t at the right end of the current word if its last letter is s (e.g., *tss* could become *tsst*).

a. Use the forward process to derive all possible words

that can be obtained in three steps by repeatedly applying the above rules in any order to the initial word *s*.

b. Apply the backward process one step to the word *tst*. Specifically, list all the words for which an application of one of the above rules would result in *tst*.

c. Prove that "If *s* then *tst*."

d. Prove that "If *s* then *ttst*."

2.20. Prove that if XYZ is an isosceles right triangle, then the hypotenuse is $\sqrt{2}$ times as long as one of the legs.

2.21. Prove that if the right triangle XYZ of Figure 2.1 on page 12 is isosceles, then the area of the triangle is $\frac{z^2}{4}$.

2.22. Prove that the triangles ABC and ADE below are similar.

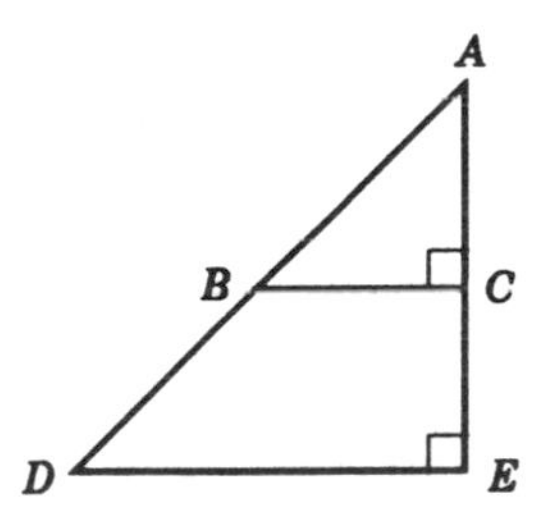

2.23. Prove that the statement in Exercise 2.11(b) is true.

THREE

ON DEFINITIONS AND MATHEMATICAL TERMINOLOGY

In the previous chapter you learned the forward–backward method and saw the importance of formulating and answering the key question. One of the simplest yet most effective ways of answering a key question is through the use of a *definition*, as will be explained in this chapter. You will also learn some of the "vocabulary" of the language of mathematics.

DEFINITIONS

A **definition** in mathematics is an agreement, by all parties concerned, as to the meaning of a particular term. You have already come across a definition in Chapter 1. There we defined what it means for the statement "A implies B" to be true—we agreed that it is true in all cases except when A is true and B is false. Nothing says that you must accept this definition as being correct. If you choose not to, then we will be unable to communicate regarding this particular idea.

Definitions are not made randomly. Usually they are motivated by a mathematical concept that occurs repeatedly. A definition can be viewed as an abbreviation that is agreed on for a particular concept. Take, for example, the notion of "a positive integer greater than one that is not divisible by any positive integer other than one and itself." This type of number is abbreviated (or defined) as a "prime." Surely it is easier to say "prime" than "a positive integer greater than one ...," especially if the concept comes up frequently. Several other examples of definitions would be:

Definition 1. An integer n **divides** an integer m (written $n|m$) if $m = kn$ for some integer k.

Definition 2. A positive integer $p > 1$ is **prime** if the only positive integers that divide p are 1 and p.

Definition 3. A triangle is **isosceles** if two of its sides have equal length.

Definition 4. Two pairs of real numbers (x_1, y_1) and (x_2, y_2) are **equal** if $x_1 = x_2$ and $y_1 = y_2$.

Definition 5. An integer n is **even** if and only if the remainder on dividing n by 2 is 0.

Definition 6. An integer n is **odd** if and only if $n = 2k + 1$ for some integer k.

Definition 7. A real number r is **rational** if and only if r can be expressed as the ratio of two integers p and q in which the denominator q is not 0.

Definition 8. Two statements A and B are **equivalent** if and only if "A implies B" and "B implies A."

Definition 9. The statement A **AND** B (written $A \wedge B$) is true if and only if A is true and B is true.

Definition 10. The statement A **OR** B (written $A \vee B$) is true in all cases except when A is false and B is false.

Observe that the words "if and only if" have been used in some of the definitions, but often, "if" tends to be used instead of "if and only if." Some terms, such as "set" and "point," are left undefined. One could possibly try to define a set as a collection of objects, but to do so is impractical because the concept of an "object" is too vague; one would then be led to ask for the definition of an "object," and so on, and so on. Such philosophical issues are beyond the scope of this book.

In the proof of Example 1, a definition was already used to answer a key question. Recall the very first one, which was "How can I show that a triangle is isosceles?" Using Definition 3, in order to show that a triangle is isosceles, one shows that two of its sides have equal length. Definitions are equally useful in the forward process. For instance, if you know that an integer n is odd, then by Definition 6 you would know that $n = 2k + 1$ for some integer k. Using definitions to work forward and backward is a common occurrence in proofs.

It is often the case that there seem to be two possible definitions for the same concept. Take, for example, the notion of an even integer introduced in Definition 5. A second plausible definition for an even integer is "an integer that can be expressed as two times some other integer." Of course there can be only one definition for a particular concept so, when more possibilities exist, how do you select the definition and what happens to the other alternatives?

Since a definition is simply something agreed on, any one of the alternatives can be agreed on as the definition. Once the definition has been chosen, it would be advisable to establish the "equivalence" of the definition and the alternatives. For the case of an even integer, this would be accomplished by using Definition 5 and its alternative to create the statements:

A: "n is an integer whose remainder on dividing by 2 is 0."

B: "n is an integer that can be expressed as two times some integer."

To establish the fact that the definition is equivalent to the alternative, you must show that "A implies B" and "B implies A" (see Definition 8). Then you would know that if A is true (i.e.,

n is an integer whose remainder on dividing by 2 is 0), then B is true (i.e., n is an integer that can be expressed as two times some other integer). Moreover, if B is true then A is true too.

The statement that A is equivalent to B is often written "A is true if and only if B is true," or, more simply, "A if and only if B." In mathematical notation one would write "A iff B" or "$A \Leftrightarrow B$." Whenever you need to show that "A if and only if B," you must show that "A implies B" and "B implies A."

It is quite useful to be able to establish that a definition is equivalent to an alternative. To see why, suppose that in some proof you derive the key question, "How can I show that an integer is even?" As a result of having obtained the equivalence of the two concepts, you now have two possible answers at your fingertips. One is obtained directly from the definition: show that its remainder on dividing by 2 is 0; the second answer comes from the alternative: show that the integer can be expressed as two times some other integer. Similarly, in the forward process, if you know that n is an even integer, then you would have two possible statements that are true as a result of this: the original definition and the alternative. While the ability to answer a key question (or to go forward) in more than one way can be a hindrance, as was the case in Example 1, it can also be advantageous, as is shown in the next example.

EXAMPLE 2.

Proposition. If n is an even integer then n^2 is an even integer.

Analysis of Proof. Proceeding by the forward–backward method, you are led immediately to the key question, "How can I show that an integer (namely, n^2) is even?" By choosing the alternative over the definition, you can answer this question by showing that

B1: n^2 can be expressed as two times some other integer.

The only question is which integer. The answer comes from the forward process.

Since n is an even integer, using the alternative, n can be expressed as two times some other integer, say k, that is,

A1: $n = 2k$.

Squaring both sides of $A1$ and rewriting by algebra yields

A2: $n^2 = (n)(n) = (2k)(2k) = 4k^2 = 2(2k^2)$.

Thus, it has been shown that n^2 can be written as two times some other integer, that integer being $2k^2$, and this completes the proof. Of course this problem could also have been solved by using Definition 5, but it is harder that way.

Proof of Example 2. Since n is an even integer, there is an integer k for which $n = 2k$. Consequently $n^2 = (2k)^2 = 2(2k^2)$, and so n^2 is an even integer. ∎

A definition is one common method for working forward and for answering certain key questions. The more statements you can show are equivalent to the definition, the more ammunition you will have available for the forward and backward processes; however, a large number of equivalent statements can also make it difficult to know exactly which one to use.

Notational difficulties can sometimes occur when using definitions in the forward and backward processes. The reason is that the definition uses one set of symbols and notation while the specific problem under consideration uses a second set of symbols and notation. When these two sets of symbols are completely distinct from each other, generally no confusion will arise; however, great care is needed when these two sets involve *overlapping notation*—that is, when the same symbol is used in both sets.

To illustrate, recall Definition 1 above:

Definition 1. An integer n **divides** an integer m (written $n|m$) if $m = kn$ for some integer k.

Suppose the conclusion of the problem you are working with is:

B: The integer p divides the integer q.

The key question associated with B is: "How can I show that one integer (namely, p) divides another integer (namely, q)?" Since the symbols p and q in the specific problem (B) are completely distinct from those in the definition, you should have no trouble using the definition to obtain the following answer to the key question:

B1: $q = kp$ for some integer k.

Observe that $B1$ was obtained by ***matching up the notation*** in Definition 1 with that in B. That is, the symbol q in B was "matched" to the symbol m in the definition; the symbol p in B was "matched" to the symbol n in the definition. $B1$ was then obtained by replacing m everywhere in the definition with q and n everywhere with p.

This process of matching up the notation when using a definition is similar to a process you have seen when working with functions. To illustrate, suppose f is a function of one variable defined by $f(x) = x(x + 1)$. To write $f(a + b)$, you "match" x to $(a + b)$, that is, you replace x everywhere by $(a + b)$ to obtain $f(a + b) = (a + b)(a + b + 1)$.

To see how notational problems can arise when using definitions, suppose the last statement in the backward process of the problem you are working with is:

B2: The integer k divides the integer n.

Once again, the key question associated with B is: "How can I show that one integer (namely, k) divides another integer (namely, n)?" A difficulty in applying Definition 1 to answer this question arises because of overlapping notation—the symbols k and n appear both in the conclusion ($B2$) and in the definition *but they are used differently in each case*. Observe in this case how complicated it is to match up the notation in $B2$ with that in the definition.

When overlapping notation occurs, you can avoid notational errors by first rewriting the definition using a *new* set of symbols so that there is no overlapping notation with the specific problem under consideration. Then when you apply the definition to the specific problem, the matching up of notation will be clear. For the example above, you could rewrite the definition as follows so that it contains no overlapping notation with $B2$:

Definition 1. An integer a **divides** an integer b (written $a|b$) if $b = ca$ for some integer c.

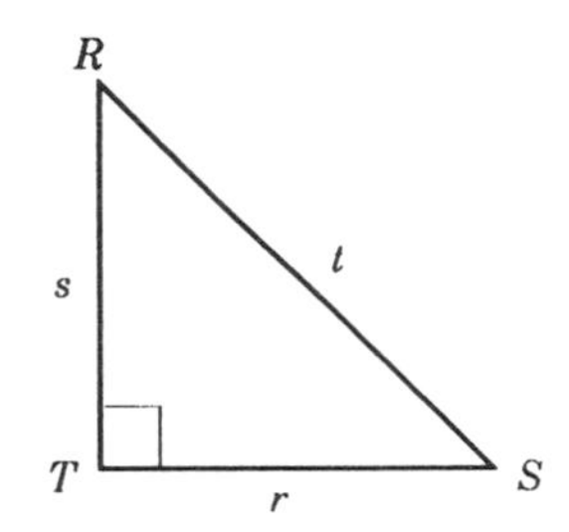

Figure 3.1 The Right Triangle RST

Now when you apply this definition to answer the key question associated with $B2$, you should have no trouble matching up the notation correctly (match k to a and n to b) to obtain the following answer:

B3: $n = ck$ for some integer c.

It is critical to learn how to apply a definition correctly in the forward and backward processes because, if you make a mistake, the rest of the proof will be incorrect. With practice, you will find that there really is no need to rewrite a definition. This is because you will be able to match up notation correctly even when there is overlapping notation. Until you reach that point, however, it is advisable to rewrite the definition.

USING PREVIOUS KNOWLEDGE

Just as a definition can be used in the forward and backward processes, so can a (previously proven) if/then statement, as will be shown in the next example.

EXAMPLE 3.

Proposition. If the right triangle RST with sides of lengths r and s, and hypotenuse of length t satisfies $t = \sqrt{2rs}$, then the triangle RST is isosceles (see Figure 3.1).

Analysis of Proof. The forward–backward method gives rise to the key question, "How can I show that a triangle (namely, RST) is isosceles?" One answer is to use Definition 3, but a second answer is also provided by the conclusion of Example 1, which states that the triangle XYZ is isosceles. Perhaps the current triangle RST is also isosceles for the same reason as triangle XYZ is. In order to find out, it is necessary to see if RST also satisfies the hypothesis of Example 1, as did triangle XYZ, for then RST will also satisfy the conclusion, and hence be isosceles.

In verifying the hypothesis of Example 1 for the triangle RST, it is first necessary to match up the current notation with that of Example 1 (just as is done when applying a definition). In this case, the corresponding lengths are $x = r$, $y = s$, and $z = t$. Thus, to check the hypothesis of Example 1 for the current problem, you must show that

B1: the area of triangle RST equals $\frac{t^2}{4}$,

or equivalently, since the area of triangle RST is $rs/2$, you must show that

B2: $\frac{rs}{2} = \frac{t^2}{4}$.

The fact that $\frac{rs}{2} = \frac{t^2}{4}$ will be established by working forward from the current hypothesis that

A: $t = \sqrt{2rs}$.

To be specific, on squaring both sides of the equality in A and dividing by 4, one obtains

A1: $\frac{rs}{2} = \frac{t^2}{4}$,

as desired. Do not forget to observe that the hypothesis of Example 1 also requires that the triangle RST be a right triangle, which of course it is, as stated in the current hypothesis.

Notice how much more difficult it would have been to match up the notation if the current triangle had been labeled WXY with sides of length w and x, and hypotenuse of length y. This overlapping notation can (and will) arise and, when it does, you should use the same technique as in the case of definitions, that is, you should rewrite the previous proposition with a set of symbols that do not overlap with the current problem.

In the condensed proof that follows, note the complete lack of reference to the matching of notation.

Proof of Example 3. By the hypothesis, $t = \sqrt{2rs}$, so $t^2 = 2rs$, or equivalently, $\frac{t^2}{4} = \frac{rs}{2}$. Thus the area of the right triangle $RST = \frac{t^2}{4}$. As such, the hypothesis, and hence the conclusion, of Example 1 is true. Consequently, the triangle RST is isosceles. ∎

MATHEMATICAL TERMINOLOGY

In dealing with proofs, there are four terms you will often come across in mathematics: *proposition, theorem, lemma,* and *corollary.* A **proposition** is a true statement of interest that you are trying to prove. All the examples that have been presented here are propositions. Some propositions are (subjectively) considered to be extremely important, and these are referred to as **theorems.** The proof of a theorem can be very long, and it is often easier to communicate the proof in "pieces." For example, in proving the statement "A implies B," it may first be convenient to show that "A implies C," then that "C implies D," and finally that "D implies B." Each of these supporting propositions might be presented separately and would be referred to as a **lemma.** In other words, a lemma is a preliminary proposition that is to be used in the proof of a theorem. Once a theorem has been established, it is often the case that certain propositions follow almost immediately as a result of knowing that the theorem is true. These are called **corollaries.**

Just as there are certain mathematical concepts that are accepted without a formal definition, so are there certain statements that are accepted without a formal proof. These unproved statements are called **axioms.** One example of an axiom is the statement: "the shortest distance between two points is a straight line." A further discussion of axioms is beyond the scope of this book.

Associated with a statement A is the statement NOT A (sometimes written $\neg A$ or $\sim A$). The statement NOT A is true when A is false, and vice versa. More will be said about the NOT of a statement in Chapter 11.

Given two statements A and B, you have already learned the meaning of the statement "A implies B." There are many other ways of saying the statement "A implies B," for example:

1. Whenever A is true, B must also be true.
2. B follows from A.
3. B is a necessary consequence of (or condition for) A (meaning that if A is true, B is *necessarily* true also.)
4. A is sufficient for B (meaning that, if you want B to be true, it is enough to know that A is true.)
5. A only if B.

Three other statements closely related to "A implies B" are given in Table 3 below.

Table 3. Statements Related to "A implies B"

Statement	Alternate Written Form	Name of Statement
"B implies A"	$(B \Rightarrow A)$	**converse**
"NOT A implies NOT B"	$(\neg A \Rightarrow \neg B)$	**inverse**
"NOT B implies NOT A"	$(\neg B \Rightarrow \neg A)$	**contrapositive**

Table 1 can be used to determine when each of these three statements is true. For instance, the contrapositive statement, "NOT B implies NOT A," is true in all cases except when the statement to the left of the word "implies" (namely, NOT B) is true and the statement to the right of the word "implies" (namely, NOT A) is false. In other words, the contrapositive statement is true in all cases except when B is false and A is true, as shown in Table 4.

Table 4. The Truth Table for "NOT B Implies NOT A"

A	B	NOT B	NOT A	$A \Rightarrow B$	NOT $B \Rightarrow$ NOT A
True	True	False	False	True	True
True	False	True	False	False	False
False	True	False	True	True	True
False	False	True	True	True	True

Note, from Table 4, that the statement "NOT B implies NOT A" is true under the same conditions as "A implies B," that is, in all cases except when A is true and B is false. This observation gives rise to a new proof technique known as the *contrapositive method* that will be described in Chapter 10. Truth tables similar to Table 4 can be derived for the converse and inverse statements and are left as exercises.

SUMMARY

This chapter has shown how definitions and previous propositions are used in the forward–backward method. A definition is used in the backward process to answer a key question, or in the forward process to derive new statements. To use a definition, you must correctly match the notation of the current problem to that of the definition. If there is overlapping notation between the two, you may want to rewrite the definition using a different set of symbols.

To use a previous proposition to prove a new one, follow these steps:

1. Find a previous proposition whose conclusion is precisely the same as the one under consideration (except possibly for the notation).
2. Match up the current notation to that of the previous

proposition. (If there is overlapping notation, you may first want to rewrite the previous one so that there is no overlapping notation.)

3. Verify that the hypothesis of the previous proposition is true for the current one. (The previous hypothesis written in terms of the notation of the current problem becomes the new statement to be proved in the *backward* process.)
4. State that the conclusion of the previous proposition is true for the current problem. (Stating the previous conclusion in terms of the notation of the current problem becomes the next statement in the *forward* process.)

You have also learned many of the terms used in the language of mathematics: a proposition is a true statement that you are trying to prove; a theorem is an important proposition; a lemma is a preliminary proposition that is to be used in the proof of a theorem; and a corollary is a proposition that follows from a theorem. Now it is time to learn more proof techniques.

EXERCISES

Note: All proofs should contain an analysis of proof as well as a condensed version. Where necessary, use the definitions on page 30.

3.1. For each of the following conclusions, pose a key question. Then use a definition (1) to answer the question abstractly and (2) to apply the answer to the specific problem.

a. If n is an odd integer, then n^2 is an odd integer.
b. If s and t are rational numbers with $t \neq 0$, then $\frac{s}{t}$ is rational.
c. Suppose that a, b, c, d, e, and f are real numbers with the property that $ad - bc \neq 0$. If (x_1, y_1) and (x_2, y_2) are pairs of real numbers satisfying:

$$ax_1 + by_1 = e, \qquad cx_1 + dy_1 = f,$$
$$ax_2 + by_2 = e, \qquad cx_2 + dy_2 = f,$$

then (x_1, y_1) equals (x_2, y_2).

d. If n is an integer greater than 1 for which $2^n - 1$ is prime, then n is prime.

e. If $(n-1)$, n, and $(n+1)$ are three consecutive integers, then 9 divides the sum of their cubes.

3.2. For each of the following statements, obtain a new statement in the backward process by using a definition. Be sure to rewrite the definitions so that no overlapping notation occurs.

a. m divides n.

b. (y_1, z_1) equals (z_2, y_2).

c. k is odd.

d. p^2 is rational.

3.3. For each of the following hypotheses, use a definition to work forward one step.

a. If n is an odd integer then n^2 is an odd integer.

b. If s and t are rational numbers with $t \neq 0$, then $\frac{s}{t}$ is rational.

c. If triangle RST is equilateral then the area of RST is $\frac{\sqrt{3}}{4}$ times the square of the length of a side.

d. If the right triangle XYZ of Figure 2.1 on page 12 satisfies $\sin(X) = \cos(X)$, then triangle XYZ is isosceles.

e. If a, b, and c are integers for which $a|b$ and $b|c$, then $a|c$.

3.4. Use a definition to work forward from each of the following statements. Be sure to rewrite the definitions so that no overlapping notation occurs.

a. The set T is a subset of the set $S \cup R$. (See Definition 14 on page 55.)

b. The real number t is an upper bound for a set T of real numbers. (See Definition 16 on page 85.)

c. The function $f + g$ of one variable is continuous at the point x, where $f + g$ is the function whose value at any point x is $f(x) + g(x)$. (See Definition 20 on page 173.)

3.5. Write truth tables for the following statements.

a. The converse of "A implies B."

b. The inverse of "A implies B." How are (a) and (b) related?

c. A OR B.

d. A AND B.

e. A AND NOT B.

f. (NOT A) OR B. How is (f) related to "A implies B"?

3.6. Suppose that A, B, and C are statements. Is the statement "A implies (B implies C)" equivalent to "(A AND B) implies C"? Why or why not? Explain.

3.7. For each of the following propositions, write the converse, inverse, and contrapositive statements.

a. If n is an integer for which n^2 is even then n is even.

b. If r is a real number such that $r^2 = 2$, then r is not rational.

c. If the quadrilateral $ABCD$ is a parallelogram with one right angle, then $ABCD$ is a rectangle.

d. Suppose that t is an angle between 0 and π. If t satisfies $\sin(t) = \cos(t)$ then $t = \frac{\pi}{4}$.

3.8. Prove that if m and n are even integers then $n+m$ is even.

3.9. Prove that if n is an odd integer then n^2 is an odd integer.

3.10. Use the proposition in Exercise 3.9 to prove that if a and b are consecutive integers, then $(a+b)^2$ is an odd integer.

3.11. Prove that if n is an odd integer and m is an odd integer, then mn is an odd integer.

3.12. Prove that if "A implies B," "B implies C," and "C implies A" then A is equivalent to B and A is equivalent to C.

3.13. Prove that if "A implies B" and "B implies C," then it follows that "A implies C."

3.14. Explain precisely how Example 2 on page 32 could be used to prove that if a and b are even integers then $(a+b)^2$ is an even integer. Specifically, match up the notation between this proposition and that in Example 2; then indicate what has to be verified before you can claim that the conclusion of Example 2 applies to the current proposition.

3.15. Suppose that a definition is given in the form of a statement A together with three possible alternative definitions, say B, C, and D.

a. How many different implications would you have to prove in order to show that A is equivalent to each of the three alternatives?

b. How many proofs would it require to show that "A implies B," "B implies C," "C implies D," and "D implies A"?

c. Explain why the approach in part (b) is sufficient to establish that each of the alternatives is equivalent to the original definition (and to each other).

3.16. Suppose you have already proved the proposition:

If a and b are nonnegative real numbers, then $(a+b)/2 \geq \sqrt{ab}$.

a. Explain precisely how this proposition could be used to prove that if a and b are real numbers satisfying the property that $b \geq 2|a|$, then $b \geq \sqrt{b^2 - 4a^2}$. Be careful how you match up notation.

b. Use the above proposition and part (a) to prove that if a and b are real numbers for which $a < 0$ and $b \geq 2|a|$, then one of the roots of the equation $ax^2 + bx + a = 0$ is $\leq -\frac{b}{a}$.

3.17. Prove that if the right triangle UVW with sides of lengths

u and v, and hypotenuse of length w satisfies $\sin(U) = \sqrt{\frac{u}{2v}}$, then the triangle UVW is isosceles, by:

a. Using the definition of an isosceles triangle.

b. Verifying the hypothesis of Example 1 on page 11.

c. Verifying the hypothesis of Example 3 on page 35.

FOUR

QUANTIFIERS I: THE CONSTRUCTION METHOD

In the previous chapter you saw that a definition could successfully be used to answer a key question. The next five chapters provide several other techniques for formulating and answering a key question that arises when B has a special form.

Two particular forms of B appear repeatedly throughout all branches of mathematics. They can always be identified by certain *key words* that appear in the statement. The first one has the words *there is (there are, there exists)*; the second one has *for all (for each, for every, for any)*. These two groups of words are referred to collectively as **quantifiers**, and each one gives rise to its own proof technique. The remainder of this chapter deals with the **existential quantifier** "there is." The **universal quantifier** "for all" is discussed in the next chapter.

WORKING WITH THE QUANTIFIER "THERE IS"

The quantifier "there is" arises quite naturally in many mathematical statements. Recall Definition 7 for a rational number as being a real number that can be expressed as the ratio of two integers in which the denominator is not zero. This definition could just as well have been written using the quantifier "there are."

Definition 11. A real number r is **rational** if and only if there are integers p and q with $q \neq 0$ such that $r = \frac{p}{q}$.

Another such example arises from the alternative definition of an even integer, that being, an integer that can be expressed as the product of two times some integer. Using a quantifier to express this statement, one obtains:

Definition 12. An integer n is **even** if and only if there is an integer k such that $n = 2k$.

It is important to observe that the quantifier "there is" allows for the possibility of more than one such object, as is shown in the next definition.

Definition 13. An integer n is a **square** if there is an integer k such that $n = k^2$.

Note that if a nonzero integer n (say, for example, $n = 9$) is square, then there are two values of k that satisfy $n = k^2$ (in this case, $k = 3$ or -3). More will be said in Chapter 12 about the issue of *uniqueness* (i.e., the existence of only one such object).

There are many other instances in which an existential quantifier can and will be used but, from the examples above, you can see that such statements always have the same basic structure. Each time the quantifier "there is," "there are," or "there exists" appears, the statement will have the following *standard form*:

There is an "object" with a "certain property" such that "something happens."

The words in quotation marks depend on the particular statement under consideration. You must learn to read, to identify, and to write each of the three components. Consider these examples.

1. There is an integer $x > 2$ such that $(x^2 - 5x + 6) = 0$.

   Object: integer x.
   Certain property: $x > 2$.
   Something happens: $(x^2 - 5x + 6) = 0$.

2. There are real numbers x and y both > 0 such that $(2x + 3y) = 8$ and $(5x - y) = 3$.

   Object: real numbers x and y.
   Certain property: $x > 0$, $y > 0$.
   Something happens: $(2x + 3y) = 8$ and $(5x - y) = 3$.

Mathematicians often use the symbol $\exists$ to abbreviate the words "there is" ("there are," etc.) and the symbol $\ni$ for the words "such that" ("for which," etc.). The use of the symbols is illustrated in the next example.

3. $\exists$ an angle $t \ni \cos(t) = t$.

   Object: angle t.
   Certain property: none.
   Something happens: $\cos(t) = t$.

Observe that the words "such that" (or equivalent words like "for which") always precede the something that happens. Practice is needed to become fluent at reading and writing these statements.

HOW TO USE THE CONSTRUCTION METHOD

During the *backward* process, if you ever come across a statement having the quantifier "there is" in the standard form:

B: There is an "object" with a "certain property" such that "something happens,"

then one way in which you can proceed to show that the statement is true is through the **construction method**. The idea is to construct (guess, produce, devise an algorithm to produce, etc.) the desired object. *However, you should realize that the construction of the object does not, by itself, constitute the proof*; rather, the proof consists of showing that the object you constructed is in fact the correct one, that is, that the object has the certain property and satisfies the something that happens.

How you actually construct the desired object is not at all clear. Sometimes it will be by trial and error; sometimes an algorithm can be designed to produce the desired object—it all depends on the particular problem. In almost all cases, the information in statement A will be used to help accomplish the task. Indeed, the appearance of the quantifier "there is" in the backward process strongly suggests turning to the forward process to produce the desired object. The construction method was used subtly in Example 2, but another example will clarify the process.

EXAMPLE 4.

Proposition. If a, b, c, d, e, and f are real numbers such that $(ad-bc) \neq 0$, then the two equations $(ax+by) = e$ and $(cx+dy) = f$ can be solved for x and y.

Analysis of proof. On starting the backward process, you should recognize that the statement B has the form discussed above, even though the quantifier "there are" does not appear explicitly. Observe that the statement B can be rewritten to contain the quantifier explicitly:

B: There are real numbers x and y such that $(ax + by) = e$ and $(cx + dy) = f$.

Statements containing "hidden" quantifiers occur frequently in problems and you should watch for them.

Proceeding with the construction method, the first step is to identify the objects, the certain property, and the something that happens. In this case one has:

Objects: real numbers x and y.
Certain property: none.
Something happens: $(ax + by) = e$ and $(cx + dy) = f$.

The next step is to construct these real numbers. You should turn to the forward process to do so. If you are able to "guess" that $x = (de-bf)/(ad-bc)$ and $y = (af-ce)/(ad-bc)$, then you are very fortunate. (Observe that, by guessing these values for x and y, you have used the information in A since the denominators are not 0.) However, recall that constructing these objects does *not* constitute the proof; you must still show that these objects satisfy the certain property and the something that happens. In this case that means you must show that, for the specific values of x and y constructed above, $(ax + by) = e$ and $(cx + dy) = f$. In doing this proof, you could write:

Noting that $ad - bc \neq 0$, construct

A1: $x = (de - bf)/(ad - bc)$ and $y = (af - ce)/(ad - bc)$.

It must be shown that

B1: $(ax + by) = e$ and $(cx + dy) = f$.

The remainder of this proof consists of working forward from $A1$ using algebra to show that $B1$ is true. The details are straightforward and will not be given here.

While this "guess and check" approach is perfectly acceptable for producing the desired x and y, it is not very informative as to *how* these particular values were produced. A more instructive proof would be desirable. For example, to obtain the values for x and y, you could start with the two equations $(ax + by) = e$ and $(cx + dy) = f$. On multiplying the first equation by d and the second one by b and then subtracting the second one from the first one, you obtain $(ad - bc)x = (de - bf)$. If you then use the information in A, it is possible to divide this last equation by $(ad - bc)$ since, by hypothesis, this number is not 0, thus obtaining $x = (de - bf)/(ad - bc)$. A similar process can be used to obtain $y = (af - ce)/(ad - bc)$. Observe that one can work backward from the desired properties to construct the object.

Even with all this added explanation as to how the objects are constructed, the mere construction of the objects does not constitute the proof. You must still show that, for these values of x and y, $(ax + by) = e$ and $(cx + dy) = f$.

Proof of Example 4. On multiplying the equation $(ax + by) = e$ by d, and the equation $(cx + dy) = f$ by b, and then subtracting the two equations one obtains $(ad - bc)x = (de - bf)$. From the hypothesis, $(ad - bc) \neq 0$, and so dividing by $(ad - bc)$ yields $x = (de - bf)/(ad - bc)$. A similar argument shows that $y = (af - ce)/(ad - bc)$, and it is not hard to check that, for these values of x and y, $(ax + by) = e$ and $(cx + dy) = f$. ∎

In the condensed proof above, observe that the author glosses over the fact that the constructed objects do satisfy the needed properties. When reading such proofs, you will have to fill in those details yourself; when writing such proofs, be sure to include those details.

SUMMARY

The construction method is not the only technique available for dealing with statements in the backward process that have the quantifier "there is," but it often works and should be considered seriously. To be successful with the construction method, you should:

1. Identify the object, the certain property, and the something that happens in the backward statement containing the quantifier "there is."
2. Turn to the forward process and use the hypothesis together with your creative ability to construct the desired object.(The actual construction of the object becomes the new statement it the *forward* process.)
3. Establish that the object you have constructed does satisfy the certain property and the something that happens. (The statement that the constructed object does satisfy the desired properties becomes the new statement in the *backward* process.)

EXERCISES

Note: All proofs should contain an analysis of proof as well as a condensed version.

4.1. For each of the following statements, identify the objects, the certain property, and the something that happens.

a. In the Himalayas, there is a mountain over 20,000 feet high that is taller than every other mountain in the world.

b. There exists an integer x satisfying $x^2 - \frac{5x}{2} + \frac{3}{2} = 0$.

c. Through a point P not on a line ℓ, there is a line ℓ' through P parallel to ℓ.

d. There exists an angle t between 0 and $\frac{\pi}{2}$ such that $\sin(t) = \cos(t)$.

e. Between the two real numbers x and y, there are distinct rational numbers r and s for which $|r - s| < 0.001$.

4.2. Suppose each of the statements in Exercise 4.1 [except for part (a)] were the conclusion of a proposition. Explain precisely how you would apply the construction method to do the proof. (You need not actually do the proof.)

4.3. Rewrite the following statements using the symbols $\exists$ and $\ni$.

a. A triangle XYZ is isosceles if two of its sides have equal length.

b. Given an angle t, one can find an angle t' whose tangent is larger than that of t.

c. At a party of n (≥ 2) people, at least two of the people have the same number of friends.

d. A polynomial of degree n, say $p(x)$, has exactly n complex roots, say $r_1, \ldots, r_n$, for which $p(r_1) = \ldots = p(r_n) = 0$.

4.4. Would you use the construction method to prove each of the following propositions? Why or why not? Explain.

a. If ABC is a right triangle, then ABC has an angle less than or equal to 45 degrees.

b. If there are real numbers x and y that lie on the two lines $y = m_1x + b_1$ and $y = m_2x + b_2$ then $m_1 \neq m_2$.

c. If a and b are real numbers for which $f(x) = ax^2 + bx + a$ has a rational root whose numerator is 1, then $f(x)$ has an integer root.

4.5. Prove each of the following statements using the construction method. Is the constructed object unique?

a. There is an integer x such that $x^2 - \frac{5x}{2} + \frac{3}{2} = 0$.

b. There is a real number x such that $x^2 - \frac{5x}{2} + \frac{3}{2} = 0$.

4.6. Prove that if a and b are real numbers with $b^2 \leq 4a$, then the two circles $x^2 + y^2 = a$ and $(x - b)^2 + y^2 = a$ intersect.

4.7. Prove that if a, b, and c are integers for which $a|b$ and $b|c$, then $a|c$.

4.8. Prove that if m and n are two consecutive integers, then 4 divides $m^2 + n^2 - 1$.

4.9. Prove that if s and t are rational numbers and $t \neq 0$, then $\frac{s}{t}$ is a rational number.

4.10. Explain what is wrong with the following condensed proof of the proposition: If a, b, and c are real numbers for which the function $ax^2 + bx + c$ has a rational root, then the function $cx^2 + bx + a$ has a rational root.

Proof. Since the function $ax^2 + bx + c$ has a rational root, there are integers p and q with $q \neq 0$ such that $a(\frac{p}{q})^2 + b(\frac{p}{q}) + c = 0$. On multiplying through by q^2 and then dividing by p^2, it follows that $x = (\frac{q}{p})$ is a rational root of $cx^2 + bx + a$. ∎

FIVE

QUANTIFIERS II: THE CHOOSE METHOD

This chapter develops the **choose method**, a proof technique for dealing with statements in the *backward* process that contain the quantifier "for all." Such statements arise quite naturally in many mathematical areas, one of which is set theory. Some time will be devoted to this topic right now because it demonstrates the use of the quantifier "for all."

WORKING WITH THE QUANTIFIER "FOR ALL"

A **set** is nothing more than a collection of items. For example, the numbers 1, 4, and 7 can be thought of as a collection of items and hence they form a set. Each of the individual items is called a *member* or *element* of the set and each member of the set is said to *be in* or *belong to* the set. The set is usually denoted by enclosing the list of its members (separated by commas) in braces. Thus, the set consisting of the numbers 1, 4, and 7 would be written

$\{1, 4, 7\}$.

To indicate that the number 4 belongs to this set, mathematicians would write

$$4 \in \{1, 4, 7\},$$

where the symbol $\in$ stands for the words "is a member of." Similarly, to indicate that 2 is not a member of $\{1, 4, 7\}$, one would write

$$2 \notin \{1, 4, 7\}.$$

While it is certainly desirable to make a list of all the elements in a set, sometimes it is impractical to do so because the list is simply too long. For example, imagine having to write down every integer between 1 and 100,000. When a set has an infinite number of elements (such as the set of real numbers that are greater than or equal to 0), it will actually be impossible to make a complete list, even if you want to. Fortunately, there is a way to describe such "large" sets through the use of what is known as **set-builder notation.** It involves using a verbal and mathematical description for the members of the set. Consider the example of the set of all real numbers that are greater than or equal to 0. With set-builder notation one would write

$$S = \{\text{real numbers } x : x \geq 0\},$$

where the ":" stands for the words "such that." Everything following the ":" is referred to as the **defining property** of the set. The question that one always has to be able to answer is "How do I know if a particular item belongs to the set or not?" To answer such a question, you need only check if the item satisfies the defining property. If so, then it is an element of the set; otherwise it is not. For the example above, to see if the real number 3 belongs to S, simply replace x everywhere by 3 and see if the defining property is true. In this case 3 does belong to S because 3 is ≥ 0.

Sometimes part of the defining property appears to the left of the ":" as well as to the right and, when trying to determine if a particular item belongs to such a set, be sure to verify this portion of the defining property too. For example, if $T = \{\text{real numbers } x \geq 0 : (x^2 - x + 2) \geq 0\}$, then -1 does not belong to T even though it satisfies the defining property to the right of the ":". The reason is that it does not satisfy the defining property to the left of the ":" since -1 is not ≥ 0.

From a proof-theory point of view, the defining property plays exactly the same role as a definition did: it is used to answer the key question "How can I show that an item belongs to a particular set?" One answer is to check that the item satisfies the defining property.

While discussing sets, observe that it can happen that no item satisfies the defining property. Consider, for example,

$$\{\text{real numbers } x \geq 0 : (x^2 + 3x + 2) = 0\}.$$

The only real numbers for which $(x^2 + 3x + 2) = 0$ are -1 and -2. Neither of these satisfies the defining property to the left of the ":". Such a set is said to be **empty**, meaning that it has no members. The special symbol ϕ is used to denote the empty set.

To motivate the use of the quantifier "for all," observe that it is usually possible to write a set in more than one way, for example, the two sets

$$S = \{\text{real numbers } x : (x^2 - 3x + 2) \leq 0\}$$

$$T = \{\text{real numbers } x : 1 \leq x \leq 2\},$$

where $1 \leq x \leq 2$ means that $1 \leq x$ and $x \leq 2$. Surely for two sets S and T to be the same, each element of S should appear in T and vice versa. Using the quantifier "for all," a definition can now be made.

Definition 14. A set S is said to be a **subset** of a set T (written $S \subseteq T$) if and only if for each element x in S, x is in T.

Definition 15. Two sets S and T are said to be **equal** (written $S = T$) if and only if S is a subset of T and T is a subset of S.

Like any definition, these can be used to answer a key question. Definition 14 answers the question "How can I show that a set (say, S) is a subset of another set (say, T)?" by requiring you to show that for each element x in S, x is also in T. As you will see shortly, the *choose* method will enable you to show that "for each element x in S, x is also in T." Definition 15 answers the key question "How can I show that two sets (say, S and T) are equal?" by requiring you to show that S is a subset of T and T is a subset of S.

In addition to set theory, there are many other instances where the quantifier "for all" can and will be used but, from the above example, you can see that all such statements appear to have the same consistent structure. When the quantifiers "for all," "for each," "for every," or "for any" appear, the statement will have the following *standard form* (which is similar to the one you saw in the previous chapter):

> For every "object" with a "certain property," "something happens."

The words in quotation marks depend on the particular statement under consideration, and you must learn to read, to write, and to identify these three components. Consider these examples.

1. For every angle t, $\sin^2(t) + \cos^2(t) = 1$.

   Object: angle t.
   Certain property: none.
   Something happens: $\sin^2(t) + \cos^2(t) = 1$.

Mathematicians often use the symbol $\forall$ to abbreviate the words "for all" ("for each," etc.). The use of symbols is illustrated in the next example.

2. $\forall$ real numbers $y > 0$, $\exists$ a real number $x \ni 2^x = y$.

   Object: real numbers y.
   Certain property: $y > 0$.
   Something happens: $\exists$ a real number $x \ni 2^x = y$.

Observe that a comma always precedes the something that happens.

Sometimes the quantifier is "hidden," for example, the statement "the cosine of an angle strictly between 0 and $\frac{\pi}{4}$ is larger than the sine of the angle" could be phrased equally well as "for every angle t with $0 < t < \frac{\pi}{4}$, $\cos(t) > \sin(t)$." Also, some authors write the quantifier *after* the something that happens, for example: "$2^n > n^2$ for all integers $n \geq 5$." Practice is needed to become fluent at reading and writing these statements, no matter how they are presented.

USING THE CHOOSE METHOD

During the *backward* process, if you ever come across a statement having the quantifier "for all" in the standard form

> For all "objects" with a "certain property," "something happens,"

then one way in which you might be able to show that the statement is true is to make a list of all the objects having the certain property. Then, for each one, you could try to show that the something happens. When the list is finite, this might be a reasonable way to proceed. However, more often than not, this approach will not be practicable because the list is too long, or even infinite. You have already dealt with this type of obstacle in set theory where the problem was overcome by using set-builder notation to describe the set. Here, the choose method will allow you to circumvent the difficulty.

The choose method can be thought of as a proof machine that, rather than actually checking that the something happens for each and every object having the certain property, has the *capability* of doing so. If you had such a machine, then there would be no need to check the whole (possibly infinite) list because you would know that the machine could always do so. The choose method shows you how to design the inner workings of the proof machine so that it will have this capability.

To understand the mechanics of the choose method, put yourself in the role of the proof machine and keep in mind that you need to have the capability of taking any object with the certain property and concluding that, for that object, the something happens (see

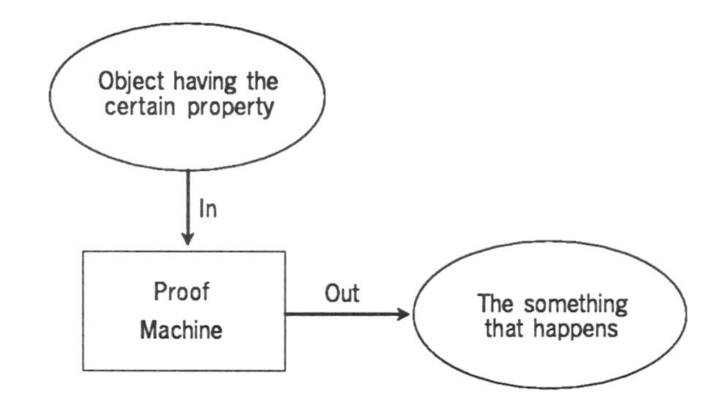

Figure 5.1 The Proof Machine for the Choose Method

Figure 5.1). As such, pretend that someone has given you one of these objects, but remember, you do not know precisely which one. All you do know is that the particular object *does have the certain property*. You must somehow be able to use that property to reach the conclusion that the something happens. This is most easily accomplished by working forward from the certain property and backward from the something that happens.

In other words, with the choose method, you *choose* one object that has the certain property. Then, by using the forward–backward method, you must conclude that, for the chosen object, the something happens. If you are successful, your proof machine will have the capability of repeating the same proof for any of the objects having the certain property.

To illustrate how the choose method is used, suppose, for example, that, in some proof, you needed to show that

B: for all real numbers x with $x^2 - 3x + 2 \leq 0$, $1 \leq x \leq 2$.

The first step is to identify, in this for-all statement, the objects (real numbers x), the certain property ($x^2 - 3x + 2 \leq 0$), and the something that happens ($1 \leq x \leq 2$). To apply the choose method, you would then choose one of these real numbers, say x', that does have the certain property [in this case, $(x')^2 - 3x' + 2 \leq 0$]. Then, by working forward from the fact that $(x')^2 - 3x' + 2 \leq 0$, you must reach the conclusion that, for x', the something happens, that is, $1 \leq x' \leq 2$.

It is important to realize that, when you apply the choose method to choose one object with the certain property, you can work forward from that information. In other words, *you have added some information to the forward process,* namely, that the chosen object satisfies the certain property. You must then show that the something happens for that chosen object, *which then becomes the new statement in the backward process.* You would indicate that the choose method is being applied to statement B in the example above by writing an analysis of the proof as follows:

Since the statement B contains the key words "for all," the choose method will be used. To that end, choose

A1: x' with the property that $(x')^2 - 3x' + 2 \leq 0$.

It will then be shown that

B1: $1 \leq x' \leq 2$.

Here, the symbol x' has been used to distinguish the chosen object from the general object x in B. Notice in $A1$ and in $B1$ that the certain property and the something that happens is written specifically for the *chosen* object, not for the general one. This is done by replacing the general object (x) everywhere in B with the chosen object (x'). In many condensed proofs, the *same* symbol is used for both the general object and the chosen one. In such cases, be careful to interpret the symbol correctly. Consider the following example.

EXAMPLE 5.

Proposition. If S and T are the two sets defined by

$$S = \{\text{real numbers } x : (x^2 - 3x + 2) \leq 0\}$$

$$T = \{\text{real numbers } x : 1 \leq x \leq 2\},$$

then $S = T$.

Analysis of proof. The forward–backward method gives rise to the key question "How can I show that two sets (namely, S and T) are equal?" Definition 15 provides the answer that you must show that

B1: S is a subset of T and T is a subset of S.

So first try to establish that

B2: S is a subset of T

and afterward, that

B3: T is a subset of S.

To show that S is a subset of T ($B2$), you obtain the key question "How can I show that a set (namely, S) is a subset of another set (namely, T)?" Using Definition 14 leads to the answer that you must show that

B4: for all elements x in S, x is in T.

This new statement, $B4$, clearly has the form described above, thus indicating that you should proceed by the choose method. To do so, first identify, in $B4$, the objects (elements x), the certain property (being in S), and the something that happens (x is in T).

To apply the choose method to $B4$, you must now choose an object having the certain property and then show that the something happens. In this case that means you should choose

A1: an element, say x, in S.

Using the fact that x is in S (i.e., that it satisfies the defining property of S) *together with the information in A*, you must show that the something happens in $B4$, that is,

B5: x is in T.

Note that you do not want to pick one specific element in S, say $\frac{3}{2}$. Also, note the double use of the symbol "x" for both the general object in $B4$ and the chosen object in $A1$.

Working backward from $B5$, you should ask the key question "How can I show that an element (namely, x) belongs to a set (namely, T)?" The answer is to show that x satisfies the defining property of T, i.e.,

B6: $1 \leq x \leq 2$.

Turning now to the forward process, you can make use of the

information in A to show that $1 \leq x \leq 2$ because you have assumed that A is true. However, there is additional information available to you. Recall that, during the backward process, you made use of the choose method, at which time you chose x to be an element in S (see $A1$). Now is the time to use the fact that x is in S—specifically, since x is in S, from the defining property of the set S, one has

A2: $x^2 - 3x + 2 \leq 0$.

Then, by factoring, one obtains

A3: $(x - 2)(x - 1) \leq 0$.

The only way that the product of $(x - 2)$ and $(x - 1)$ can be ≤ 0 is for one of them to be ≤ 0 and the other ≥ 0. In other words,

A4: Either $(x - 2) \geq 0$ and $(x - 1) \leq 0$, or else $(x - 2) \leq 0$ and $(x - 1) \geq 0$.

The first situation can never happen because, if it did, x would be ≥ 2 and x would be ≤ 1, which is impossible. Thus the second condition must happen, that is,

A5: $x \leq 2$ and $x \geq 1$.

But this is precisely the last statement obtained in the backward process ($B6$), and hence it has been shown successfully that S is a subset of T. Do not forget that you still have to show that T is a subset of S ($B3$) in order to complete the proof that $S = T$. This part, however, will be left as an exercise.

Proof of Example 5. To show that $S = T$ it will be shown that S is a subset of T and T is a subset of S. To see that S is a subset of T, let x be in S (the use of the word "let" in condensed proofs frequently indicates that the choose method has been invoked). Consequently, $(x^2 - 3x + 2) \leq 0$, and so $(x - 2)(x - 1) \leq 0$. This means that either $(x - 2) \geq 0$ and $(x - 1) \leq 0$, or else that, $(x - 2) \leq 0$ and $(x - 1) \geq 0$. The former cannot happen because, if it did, $x \geq 2$ and $x \leq 1$. Hence it must be that $x \leq 2$ and $x \geq 1$, which means that x is in T. The proof that T is a subset of S is omitted. ∎

Whenever you apply the choose method to choose an object with the certain property, you must first be sure that, indeed, there is at least one such object, for if there are none, how can you choose such an object? To illustrate, suppose you are trying to prove that

B: for all elements in $\{\text{real numbers } x \geq 0 : x^2 + 3x + 2 = 0\}$, $x^2 \leq 4$.

According to the choose method, you should choose an element $x \geq 0$ with the property that $x^2 + 3x + 2 = 0$. However, there are no such elements (because the only values for x that satisfy the equation are $x = -1$ and $x = -2$, and neither of these is ≥ 0). In this case you cannot apply the choose method; however, there is no need to do so. The reason is that when there is no object with the certain property, the associated for-all statement is automatically true, as explained in Exercise 5.5 below.

SUMMARY

Use the choose method when the last statement in the backward process contains the quantifier "for all." Proceed by

1. Identifying the object, the certain property, and the something that happens in the for-all statement.
2. Verifying that there is at least one object with the certain property.
3. Choosing an object that has the certain property. (Write the fact that the chosen object has the certain property as a new statement in the *forward* process.)
4. Show that, for this chosen object, the something happens. (Write this objective as the next statement in the *backward* process.)

The fourth step is accomplished by the forward–backward method.

That is, work forward from the fact that the chosen object (in Step 3) has the certain property and backward from the fact that this chosen object must be shown to satisfy the something that happens. In so doing, you can use the assumption that the hypothesis (A) is true, or any other statement in the forward process.

EXERCISES

Note: All proofs should contain an analysis of proof as well as a condensed version.

5.1. For each of the following definitions, identify the objects, the certain property, and the something that happens in the associated for-all statements.

a. The real number x^* is a **maximizer** of the function f if and only if, for every real number x, $f(x) \leq f(x^*)$.

b. Suppose that f and g are functions of one variable. Then g is $\geq f$ on the set S of real numbers if and only if, for every element x in S, $g(x) \geq f(x)$.

c. A real number u is an **upper bound** for a set S of real numbers if and only if, for all x in S, $x \leq u$.

d. The set C of real numbers is a **convex set** if and only if, for every element x and y in C and for every real number t between 0 and 1, $tx + (1-t)y$ is an element of C.

e. The function f of one variable is a **convex function** if and only if, for all real numbers x and y and for all real numbers $0 \leq t \leq 1$, it follows that

$$f(tx + (1-t)y) \leq tf(x) + (1-t)f(y).$$

5.2. Suppose you are trying to prove that "If f is the function defined by $f(x) = x^2 - 4x + 1$ then, for all real numbers x with $x > 4$, $f(x) > 1$." Which of the following constitute a correct application of the choose method? For those that are incorrect, explain what is wrong.

a. Choose

A1: a real number x'.

It must be shown that

B1: $f(x') > 1$.

b. Choose

A1: a real number x' with $x' > 4$.

It must be shown that

B1: $f(x') > 1$.

c. Choose

A1: a real number x with $f(x) > 1$.

It must be shown that

B1: $x > 4$.

d. Choose

A1: a real number, say 5, that is > 4.

It must be shown that

B1: $f(5) > 1$.

e. Choose

A1: a real number x with $x > 4$.

It must be shown that

B1: $f(x) > 1$.

5.3. For each of the parts in Exercise 5.1, describe how the choose method would be applied if you were trying to show that the for-all statement is true. Use a different symbol to distinguish the chosen object from the general object. For instance, for Exercise 5.1(a), to show that x^* is the maximizer of the function f, one would choose

A1: a real number, say x',

for which it must then be shown that

$$\textbf{B1:} \quad f(x') \le f(x^*).$$

5.4. Would you use the choose method to prove that "If a, b, c, and y are real numbers with the property that, for all real numbers x, $ax^2 + bx + c \le ay^2 + by + c$, then $a \le 0$?" Why or why not? Explain.

5.5. Consider the problem of showing that "For every object X with a certain property, something happens." Discuss why the approach of the choose method is the same as that of using the forward–backward method to show that "If X is an object with the certain property, then the something happens." How are the two statements in quotation marks related? Use this relationship to explain why a for-all statement in which there is no object with the certain property is necessarily true.

5.6. Using the results of Exercise 5.5, reword the following for-all statements in an equivalent "If ... then ..." form.

a. For every angle t with $0 \le t \le \frac{\pi}{2}$, $\cos(t) \ge 0$.

b. For every point P not on a line L, there is a line through P that is parallel to L.

c. $\forall$ element $x \in S$, $x \in T$ (S and T are sets).

5.7. Reword the following statements in *standard form*, using the appropriate symbols $\forall$, $\exists$, $\ni$ as necessary.

a. Some mountain is taller than every other mountain.

b. If t is an angle then $\sin(2t) = 2\sin(t)\cos(t)$. (Hint: Make use of Exercise 5.5.)

c. The square root of the product of any two nonnegative real numbers p and q is always $\le$ their sum divided by 2.

d. If x and y are real numbers such that $x < y$, then there is a rational number r such that $x < r < y$.

5.8. For the proposition and condensed proof given below, explain precisely where (i.e., in which sentence), why, and how the choose method is used. Has this technique been used correctly? Why or why not? Explain. [You may want to refer to the definition in Exercise 5.1(a).]

Proposition. If a, b, and c are real numbers for which $a < 0$ then $x^* = \frac{-b}{(2a)}$ is a maximizer of $f(x) = ax^2 + bx + c$.

Proof. Let x be a real number. If $x^* \geq x$ then $(x^* - x) \geq 0$ and $a(x^* + x) + b \geq 0$. So, $(x^* - x)[a(x^* + x) + b] \geq 0$. On multiplying the term $(x^* - x)$ through, rearranging terms, and adding c to both sides one obtains that $a(x^*)^2 + bx^* + c \geq ax^2 + bx + c$. A similar argument applies when $x^* < x$. ∎

5.9. Write an analysis of the proof that corresponds to the condensed proof given below. Indicate which proof techniques are being used and how they are applied. Fill in the details of any missing steps where appropriate.

Proposition. If m and b are real numbers with $m > 0$, and f is the function defined by $f(x) = mx + b$, then for all real numbers x and y with $x < y$, $f(x) < f(y)$.

Proof. Let x and y be real numbers with $x < y$. Then since $m > 0$, $mx < my$. On adding b to both sides, it follows that $f(x) < f(y)$, and so the proof is complete. ∎

5.10. Write an analysis of the proof that corresponds to the condensed proof given below. Indicate which proof techniques are being used and how they are applied. Fill in the details of any missing steps where appropriate.

Proposition. If $S = \{\text{real numbers } x : x(x-3) \leq 0\}$ and $T = \{\text{real numbers } x : x \geq 3\}$ then every element of T is an upper bound for the set S [see the definition in Exercise 5.1(c)].

Proof. Let t be an element of T. It will be shown that t is an upper bound for S. To that end, let x be an element of S. Consequently, $x(x-3) \leq 0$. Therefore, it must be that $x \geq 0$ and $(x-3) \leq 0$. But then $x \leq 3$, and since t belongs to T, $t \geq 3$, and so $x \leq t$. ∎

5.11. For the sets S and T of Example 5, prove that $T \subseteq S$.

5.12. Prove that if $f(x) = (x-1)^2$ and $g(x) = x+1$, then $g \geq f$ on the set $S = \{\text{real numbers } x : 0 \leq x \leq 3\}$. [See the definition in Exercise 5.1(b).]

5.13. Prove that if m and b are real numbers, and f is a function defined by $f(x) = mx + b$, then f is convex. [See the definition in Exercise 5.1(e).]

5.14. Prove that if a and b are real numbers, then the set $C = \{\text{real numbers } x : ax \leq b\}$ is a convex set. [See the definition in Exercise 5.1(d).]

SIX

QUANTIFIERS III: INDUCTION

In the previous chapter you learned how to use the choose method when the quantifier "for all" appears in the statement B. There is one special form of B containing the quantifier "for all" for which a separate proof technique known as **mathematical induction** has been developed.

HOW TO USE MATHEMATICAL INDUCTION

Induction should be considered seriously (even before the choose method) when B has the form:

For every *integer* $n \geq 1$, "something happens"

where the something that happens is some statement that depends on the integer n. An example would be the statement:

For all integers $n \geq 1$, $\sum_{k=1}^{n} k = \frac{n(n+1)}{2}$, where

$\sum_{k=1}^{n} k = 1 + \ldots + n$.

When considering induction, the key words to look for are "integer" and "≥ 1."

One way to attempt proving such statements would be to make an infinite list of statements, one for each of the integers starting from $n = 1$, and then prove each statement separately. While the first few statements on the list are usually easy to verify, the issue is how to check statement number n and beyond. For the example above, the list would be:

$$\mathbf{P(1)}: \sum_{k=1}^{1} k = \frac{1(1+1)}{2} \quad \text{or} \quad 1 = 1$$

$$\mathbf{P(2)}: \sum_{k=1}^{2} k = \frac{2(2+1)}{2} \quad \text{or} \quad 1+2 = 3$$

$$\mathbf{P(3)}: \sum_{k=1}^{3} k = \frac{3(3+1)}{2} \quad \text{or} \quad 1+2+3 = 6$$

$$\vdots$$

$$\mathbf{P(n)}: \sum_{k=1}^{n} k = \frac{n(n+1)}{2}$$

$$\mathbf{P(n+1)}: \sum_{k=1}^{n+1} k = \frac{(n+1)[(n+1)+1]}{2} = \frac{(n+1)(n+2)}{2}$$

$$\vdots$$

Induction is a clever method for proving that each of these statements in the infinite list is true. As with the choose method, induction can be thought of as a proof machine that starts with $P(1)$ and works its way progressively down the list proving each statement as it proceeds. Here is how it works.

You start the machine by verifying that $P(1)$ is true, as can be done easily for the example above. Then you feed $P(1)$ into the machine. It uses the fact that $P(1)$ is true and automatically proves that $P(2)$ is true. You then take $P(2)$ and put it into the machine. The machine uses the fact that $P(2)$ is true to reach the conclusion that $P(3)$ is true, and so on (see Figure 6.1).

Observe that, by the time the machine is going to prove that $P(n + 1)$ is true, it will already have shown that $P(n)$ is true (from the previous step). Thus, in designing the machine, *you*

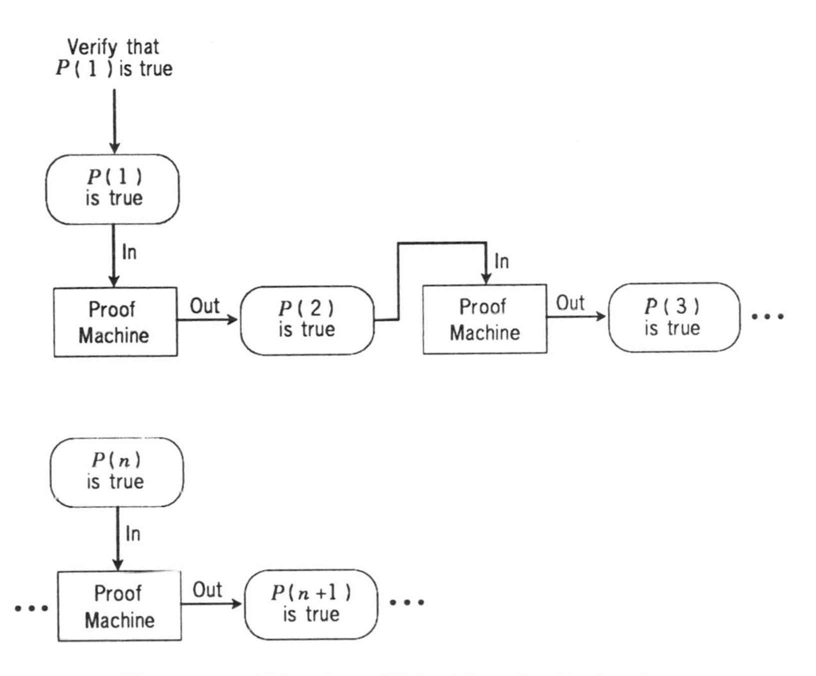

Figure 6.1 The Proof Machine for Induction

can assume that P(n) is true; your job is to make sure that P(n + 1) will also be true. Do not forget that, in order to start the whole process, *you must also verify that P(1) is true.*

The first step of an induction proof is to verify that the statement $P(1)$ is true. To do so, simply replace n everywhere in $P(n)$ by 1. To verify that the resulting statement is true usually requires only some minor rewriting.

The second step is more challenging. It requires you to reach the conclusion that $P(n+1)$ is true by using the assumption that $P(n)$ is true. There is a very standard way of doing this. Begin by writing down the statement $P(n+1)$. Since you are allowed to assume that $P(n)$ is true and you want to conclude that $P(n+1)$ is true, you should somehow try to rewrite the statement $P(n+1)$ in terms of $P(n)$—as will be illustrated in a moment—for then you will be able to make use of the assumption that $P(n)$ is true. (Making use of the assumption that $P(n)$ is true is often referred to as **using the induction hypothesis.**) On establishing that $P(n+1)$ is true, the proof will be complete.

EXAMPLE 6.

Proposition. For every integer $n \geq 1$, $\sum_{k=1}^{n} k = \frac{n(n+1)}{2}$.

Analysis of proof. When you are using the method of induction, it is helpful to write down the statement $P(n)$, in this case:

$$\mathbf{P(n)}: \ \sum_{k=1}^{n} k = \frac{n(n+1)}{2}.$$

The first step in a proof by induction is to verify $P(1)$. Replacing n everywhere by 1 in $P(n)$, you obtain

$$\mathbf{P(1)}: \ \sum_{k=1}^{1} k = \frac{1(1+1)}{2}.$$

With a small amount of rewriting, it is easy to verify this since

$$\sum_{k=1}^{1} k = 1 = \frac{1(1+1)}{2}.$$

This step is often so easy that it is virtually omitted in the condensed proof simply by saying "The statement is clearly true for $n = 1$."

The second step is more involved. You must use the assumption that $P(n)$ is true to reach the conclusion that $P(n+1)$ is true. The best way to proceed is to write down the statement $P(n+1)$ by replacing carefully n with $(n+1)$ everywhere in $P(n)$, and rewriting a bit, if necessary. In this case

$$\begin{aligned} \mathbf{P(n+1)}: \sum_{k=1}^{n+1} k &= \frac{(n+1)[(n+1)+1]}{2} \\ &= \frac{(n+1)(n+2)}{2}. \end{aligned}$$

To reach the conclusion that $P(n+1)$ is true, begin with the left side of the equality in $P(n+1)$ and try to make it look like the right side. In so doing, you should use the information in $P(n)$ by relating the left side of the equality in $P(n+1)$ to the left side of the equality in $P(n)$. Then you will be able to use the right side of the equality in $P(n)$. In this example,

$$\mathrm{P(n+1):}\quad \sum_{k=1}^{n+1} k = \left(\sum_{k=1}^{n} k\right) + (n+1).$$

Now you can use the assumption that $P(n)$ is true by replacing

$$\sum_{k=1}^{n} k$$

with $\frac{n(n+1)}{2}$, obtaining

$$\sum_{k=1}^{n+1} k = \left[\frac{n(n+1)}{2}\right] + (n+1).$$

All that remains is a bit of algebra to rewrite $[n(n+1)/2]+(n+1)]$ as $[(n+1)(n+2)/2]$, thus obtaining the right side of the equality in $P(n+1)$. The algebraic steps are:

$$\begin{aligned}\left[\frac{n(n+1)}{2}\right] + (n+1) &= (n+1)\left(\frac{n}{2}+1\right)\\ &= \frac{(n+1)(n+2)}{2}.\end{aligned}$$

In summary,

$$\begin{aligned}
\mathbf{P(n+1)}: \sum_{k=1}^{n+1} k &= \left(\sum_{k=1}^{n} k\right) + (n+1) \\
&= \left[\frac{n(n+1)}{2}\right] + (n+1) \\
&= (n+1)\left(\frac{n}{2}+1\right) \\
&= \frac{(n+1)(n+2)}{2}.
\end{aligned}$$

Your ability to relate $P(n+1)$ to $P(n)$ so as to use the induction hypothesis that $P(n)$ is true will determine the success of the proof by induction. If you are unable to relate $P(n+1)$ to $P(n)$, then you might wish to consider a different proof technique. On the other hand, if you can relate $P(n+1)$ to $P(n)$, you will find that induction is easier to use than almost any other technique. To illustrate this fact, you will be asked, in the exercises, to reprove Example 6 *without* using induction. Compare your proof with the condensed proof below.

Proof of Example 6. The statement is clearly true for $n = 1$. Assume it is true for n (i.e., that $\sum_{k=1}^{n} k = n(n+1)/2$). Then

$$\begin{aligned}
\sum_{k=1}^{n+1} k &= \left(\sum_{k=1}^{n} k\right) + (n+1) \\
&= \left[\frac{n(n+1)}{2}\right] + (n+1) \\
&= (n+1)\left(\frac{n}{2}+1\right) \\
&= \frac{(n+1)(n+2)}{2}
\end{aligned}$$

which is $P(n+1)$. ∎

SOME VARIATIONS ON MATHEMATICAL INDUCTION

From the discussion above, you now know that part of a proof by induction has you assume that $P(n)$ is true and use that fact to show that $P(n+1)$ is true. From a notational point of view, some authors prefer to assume that $P(n-1)$ is true and use that fact to show that $P(n)$ is true. These two approaches are completely identical—either one can be used, depending on your notational preference. What is important is that you establish that if *one* of the statements on the infinite list is true, then the *next* one must also be true.

When using the method of induction, it is not necessary that the first value for n has to be 1. For instance, the statement "for all integers $n \geq 5$, $2^n > n^2$" can also be proved by induction. The only modification is that, in order to start the proof, you must verify $P(n)$ for the first given value of n. In this case, that first value is $n = 5$, so you will have to check that $2^5 > 5^2$ (which it is since $2^5 = 32$ while $5^2 = 25$). The second step of the induction proof remains the same. You would still have to show that if $P(n)$ is true (i.e., $2^n > n^2$), then $P(n+1)$ is also true [i.e., $2^{n+1} > (n+1)^2$]. In so doing, you can also use the fact that $n \geq 5$, if necessary.

Another modification to the basic induction method arises when you are having difficulty relating $P(n+1)$ to $P(n)$, or when the relationship you discover does not prove fruitful. Suppose, however, that you can relate $P(n+1)$ to $P(j)$, where $j < n$. In this case, you would like to use the fact that $P(j)$ is true but, can you assume that $P(j)$ is, in fact, true? The answer is yes. To see why, recall the analogy of the proof machine (look again at Figure 6.1), and observe that, by the time the machine has to show that $P(n+1)$ is true, it will already have established that all of the statements $P(1), \ldots, P(j), \ldots, P(n)$ are true. Thus, when trying to show that $P(n+1)$ is true, you can assume that $P(n)$ *and all preceding statements are true.* Such a proof is referred to as **generalized induction** and is illustrated in the following example.

EXAMPLE 7.

Proposition. Any integer $n \geq 2$ can be expressed as a finite product of primes (see Definition 2 on page 30).

Analysis of Proof. Applying induction, this statement is clearly true for $n = 2$. Proceeding with the second step, you can assume the statement is true for n, that is,

A1: n can be expressed as a finite product of primes, say, $p_1, \ldots, p_k$.

You must conclude that $P(n+1)$ is true, that is, that

B1: $(n+1)$ can be expressed as a finite product of primes.

You might proceed by trying to relate $P(n)$ to $P(n+1)$ as follows:

$$\begin{aligned}(n+1) &= (n)+1 \\ &= (p_1 \cdot p_2 \cdot \ldots \cdot p_k)+1\end{aligned}$$

However, you will now find it difficult to express the right side of the last equality above as a product of primes.

Instead of relating $P(n+1)$ to $P(n)$, consider the following approach. Replace $A1$ above with the assumption that $P(j)$ is true for all values of $j = 2, \ldots, n$, that is, assume that

A2: any integer between 2 and n can be expressed as a finite product of primes.

To show that $B1$ is true, that is, that $P(n+1)$ is true, consider the integer $(n+1)$. If $(n+1)$ is itself prime, then $P(n+1)$ is true. If $(n+1)$ is not prime, then it has some prime divisor, that is,

A3: there are integers p and q with p being prime such that $n+1 = pq$.

But now, q is an integer between 2 and n and so you can apply generalized induction. Specifically, you can conclude that $P(q)$ is true, that is, q can be expressed as a finite product of primes, say, $p_1, \ldots, p_k$. So,

A4: there are primes $p_1, \ldots, p_k$ such that $q = p_1 \cdot \ldots \cdot p_k$.

Combining $A3$ and $A4$, you reach the desired conclusion that

A5: $n + 1 = p(p_1\, p_2\, \ldots\, p_k)$

Therefore $n + 1$ can be expressed as a finite product of primes, and so $P(n + 1)$ is true, thus completing the proof.

Proof of Example 7. The statement is clearly true for $n = 2$. Now assume the statement is true for all integers j between 2 and n. That is, that any integer j between 2 and n can be expressed as a finite product of primes. If $(n + 1)$ is prime, the statement is true for $(n + 1)$; otherwise, $(n + 1)$ has a prime divisor, say p. So there is an integer q with $2 \le q \le n$ such that $(n + 1) = pq$. But by the induction hypothesis, q can be expressed as a finite product of primes and therefore, so can $(n + 1)$. ∎

SUMMARY

Use induction when the statement you are trying to prove has the form: "For every integer n greater than or equal to n_0, $P(n)$," [where $P(n)$ is some statement that depends on n]. To apply the method of induction,

1. Verify that the statement $P(n)$ is true for n_0. [To do this, replace n everywhere in $P(n)$ by n_0, rewrite the resulting statement, and try to establish that it is true.]
2. Assume that $P(n)$ is true.
3. Write down the statement $P(n + 1)$ by replacing n everywhere in $P(n)$ with $(n + 1)$. [Some rewriting may be necessary to express $P(n + 1)$ in a clean form.]
4. Reach the conclusion that $P(n + 1)$ is true. [To do this, relate $P(n + 1)$ to $P(n)$ and then use the fact that $P(n)$ is true. The key to using induction rests in your ability to relate $P(n + 1)$ to $P(n)$.]

It is important to realize that induction does not help you to discover the correct *form* of the statement $P(n)$. Induction only verifies that a given statement $P(n)$ is true for all integers n greater than or equal to some initial one.

EXERCISES

Note: Proofs in this chapter need not contain an analysis of proof.

6.1. For which of the following statements would induction be applicable? When it is not applicable, explain why.

a. For every positive integer n, 8 divides $5^n + 2 \cdot 3^{n-1} + 1$.

b. There is an integer $n \geq 0$ such that $2n > n^2$.

c. For every integer $n \geq 1$, $1(1!) + \ldots + n(n!) = (n+1)! - 1$. (Recall that $n! = n(n-1)\ldots 1$.)

d. For every integer $n \geq 4$, $n! > n^2$.

e. For every real number $n \geq 1$, $n^2 \geq n$.

6.2. Upon learning about the method of induction, a student said, "I don't understand something. After showing that the statement is true for $n = 1$, you want me to assume that $P(n)$ is true and to show that $P(n+1)$ is true. How can I *assume* that $P(n)$ is true—after all, aren't we trying to *show* that $P(n)$ is true?" Answer this question.

6.3. a. Why and when would you want to use induction instead of the choose method?

b. Why is it not possible to use induction on statements of the form: For every "real number" with a "certain property," "something happens"?

6.4. Identify $P(n)$ in each of the following statements. Then write down $P(1)$, $P(n-1)$, and $P(n+1)$.

a. For every integer $n \geq 1$, $1(1!) + \ldots + n(n!) = (n+1)! - 1$.

b. For all integers $n \geq 1$, 6 divides $n^3 - n$.

c. A set of $n \geq 1$ elements has 2^n subsets.

6.5. Prove, by induction, that, for every integer $n \geq 1$, $1(1!) + \ldots + n(n!) = (n+1)! - 1$.

6.6. Prove, by induction, that, for every integer $p \geq 1$,

$$\sum_{k=1}^{p} k^2 = \frac{p(p+1)(2p+1)}{6}.$$

6.7. Prove, by induction, that, for every integer $n \geq 5$, $2^n > n^2$.

6.8. Prove, by induction, that, if x is a real number greater than -1, then, for every integer $n \geq 1$, $(1+x)^n \geq 1 + nx$.

6.9. Prove, by induction, that a set of $n \geq 1$ elements has 2^n subsets (including the empty set).

6.10. Prove, by induction, that for every integer $p \geq 2$,

$$\left(1 - \frac{1}{4}\right)\left(1 - \frac{1}{9}\right)\cdots\left(1 - \frac{1}{p^2}\right) = \frac{(p+1)}{(2p)}.$$

6.11. Prove, *without using induction*, that, for any integer $n \geq 1$,

$$\sum_{k=1}^{n} k = \frac{n(n+1)}{2}.$$

6.12. Prove, by induction, that, if $i^2 = -1$, then for every integer $n \geq 1$, $[\cos(x) + i\sin(x)]^n = \cos(nx) + i\sin(nx)$.

6.13. Prove that for all integers $n \geq 1$, 6 divides $n^3 - n$ by showing that (1) the statement is true for $n = 1$, and (2) if the statement is true for $(n-1)$, then it is true for n.

6.14. Prove, by induction, that, for all integers $n \geq 2$, if $x_1, x_2, \ldots, x_n$ are real numbers strictly between 0 and 1, then $(1 - x_1)(1 - x_2)\ldots(1 - x_n) > 1 - x_1 - x_2 - \ldots - x_n$.

6.15. Describe a "modified" induction procedure that could be used to prove statements of the form:

a. For every integer $\leq$ some given one, something happens.

b. For every integer, something happens.

c. For every positive odd integer, something happens.

6.16. Develop an induction-type technique that could be used for proving a statement of the form: "For every integer $n \geq 1$, and for every integer $m \geq 1$, $P(n,m)$," [where $P(n,m)$ is some statement that depends on both n and m].

6.17. What is wrong with the following proof that all horses have the same color?

Proof. Let n be the number of horses. When $n = 1$, the statement is clearly true, that is, one horse has the same color, whatever color it is. Assume that any group of n horses has the same color. Now consider a group of $(n+1)$ horses. Taking any n of them, the induction hypothesis states that they all have the same color, say brown. The only issue is the color of the remaining "uncolored" horse. Consider, therefore, any other group of n of the $(n+1)$ horses that contains the uncolored horse. Again, by the induction hypothesis, all of the horses in the new group must have the same color. Then, since all of the colored horses in this group are brown, the uncolored horse must also be brown. ∎

6.18. What is wrong with the proof given below of the statement: "If r is a real number with $|r| \leq 1$, then for all integers $n \geq 1$, $1 + r + r^2 + \ldots + r^{n-1} = \dfrac{(1 - r^n)}{(1 - r)}$."

Proof. The statement is clearly true for $n = 1$. Assume it is true for n. Then, for $n + 1$ one has

$$
\begin{aligned}
1 + \ldots + r^n &= \left[\frac{(1 - r^n)}{(1 - r)}\right] + r^n \\
&= \frac{(1 - r^n + r^n - r^{n+1})}{(1 - r)} \\
&= \frac{(1 - r^{n+1})}{(1 - r)}. \quad \blacksquare
\end{aligned}
$$

SEVEN

QUANTIFIERS IV: SPECIALIZATION

The previous three chapters illustrated how to proceed when a quantifier appears in the statement B. This chapter develops a method for exploiting quantifiers that appear in the statement A.

When the statement A contains the quantifier "there is" in the standard form:

A: There is an "object" with a "certain property" such that "something happens"

you can use this information in a straightforward way. When showing "A implies B" by the forward–backward method, you are assuming A is true. In this case that means you can assume that indeed there is an object with the certain property such that the something happens. In doing the proof, you would say: "Let X be an object with the certain property and for which the something happens ..." The existence of this object will somehow be used in the forward process to obtain the conclusion that B is true.

HOW TO USE SPECIALIZATION

The more interesting situation occurs when the statement A contains the quantifier "for all" in the standard form:

A: For all "objects" with a "certain property," "something happens."

To use the information in A, one typical method emerges and it is referred to as **specialization**. In general terms specialization works as follows. As a result of assuming A is true, you know that, for all objects with the certain property, something happens. If, at some point, you were to come across *one* of these objects that does have the certain property, then you can use the information in A by being able to conclude that, for this *particular* object, the something does indeed happen. That fact should help you to conclude that B is true. In other words, you will have specialized the statement A to one particular object having the certain property.

To illustrate the idea of specialization in a more tangible way, suppose you know that

A: All Fords made in 1988 get good gas mileage.

In this statement, you can identify the following three items:

Objects: Fords.
Certain property: made in 1988.
Something happens: get good gas mileage.

Suppose that you are interested in buying a car that gets good gas mileage, so your objective is

B: To buy a car that gets good gas mileage.

You can work forward from the information in A (by specialization) to establish B as follows. Suppose you are walking along the street one day and you see a particular Ford. Looking more carefully at the car, you verify that it was made in 1988. Recalling that statement A above is assumed to be true, you can use this information to conclude that

A1: this *particular* 1988 Ford gets good gas mileage.

In other words, you have *specialized* the for-all statement in A to one *particular* object.

If you analyze the example above in detail, you will identify the following steps associated with applying specialization to a *forward* statement of the form:

A: For all "objects" with a "certain property," "something happens."

1. Assume that A is true and identify, in the for-all statement, the objects, the certain property, and the something that happens.
2. Look for one particular object to apply specialization to. (This object often arises from the backward process.)
3. Verify that this particular object does have the certain property specified in the for-all statement in A.
4. Conclude—by writing a new statement in the forward process—that the something happens for this particular object.

The following example demonstrates the proper use of specialization in doing mathematical proofs.

Definition 16. A real number u is an **upper bound** for a set of real numbers T if for all elements t in T, $t \leq u$.

EXAMPLE 8.

Proposition. If R is a subset of a set S of real numbers and u is an upper bound for S, then u is an upper bound for R.

Analysis of proof. The forward–backward method gives rise to the key question "How can I show that a real number (namely, u) is an upper bound for a set of real numbers (namely, R)?" Definition 16 is used to answer the question. Thus it must be shown that

B1: for all elements r in R, $r \leq u$.

The appearance of the quantifier "for all" in the *backward* process suggests proceeding with the choose method, whereby one chooses

A1: an element, say r, in R

for which it must be shown that

B2: $r \leq u$.

(Note that the same symbol r has been used for the chosen object in $A1$ as for the general object in the for-all statement in $B1$.)

Turning now to the forward process, you will see how specialization is used to reach the conclusion that $r \leq u$. From the hypothesis that R is a subset of S, and by Definition 14 on page 55, you know that

A2: for each element x in R, x is in S.

Recognizing the key words "for all" *in the forward process*, you should consider using specialization. According to the discussion preceding Example 8, the first step in doing so is to identify, in $A2$, the object (element x), the certain property (being in the set R), and the something that happens (x is in the set S). Next, you must look for one particular object with which to specialize. In the backward process you came across the particular element r (see $A1$). The third step of specialization requires that you verify that the particular object (r) has the certain property in $A2$, namely, being in the set R. In fact the particular object r does have this property because r was chosen to be in the set R (see $A1$). The final step of specialization is to conclude, by writing a new statement in the forward process, that, for the particular object (r), the something happens (in $A2$). In this case, specialization allows you to conclude that

A3: r is in S.

The proof is not yet complete because the last statement in the backward process ($B2$) has not yet been reached in the forward process. To do so, continue to work forward. For example, from the hypothesis, you know that u is an upper bound for S. By Definition 16 above this means that

A4: for every element s in S, $s \leq u$.

Once again, the appearance of the quantifier "for every" in the *forward* process suggests using specialization. Accordingly, identify, in $A4$, the object (element s), the certain property (being in the set S), and the something that happens ($s \leq u$). Now look for one particular object with which to apply specialization. The same element r chosen in $A1$ serves the purpose. Next, verify that this particular r satisfies the certain property in $A4$, namely, of being in the set S. Indeed r is an element of S, as stated in $A3$. Finally, conclude that, for this particular object (r), the something in $A4$ happens, so

A5: $r \leq u$.

Since the statement "$r \leq u$" was the last one obtained in the backward process (see $B2$), the proof is now complete.

In the condensed proof of Example 8 that follows, note the lack of reference to the forward–backward, choose, and specialization methods.

Proof of Example 8. To show that u is an upper bound for R, let r be an element of R (the word "let" here indicates that the choose method has been used). By hypothesis, R is a subset of S and so r is also an element of S (here is where specialization has been used). Furthermore, by hypothesis, u is an upper bound for S, thus, every element in S is $\leq u$. In particular, r is an element of S, so $r \leq u$ (again specialization has been used). ∎

When using specialization, be careful to keep your notation and symbols in order. Doing so involves a correct "matching up of notation," similar to what you learned in Chapter 3 on using definitions. To illustrate, suppose you are going to apply specialization to a statement of the form:

A: For all objects X with a certain property, something happens.

After finding a particular object, say Y, with which to specialize, it is necessary to verify that Y satisfies the certain property in A. To do so, replace X with Y everywhere in the certain property in A and see if the resulting condition is true. Similarly, when concluding that the particular object Y satisfies the something that happens in A, again replace X everywhere with Y in the something that happens to obtain the correct statement in the forward process. (This was done when writing statements $A3$ and $A5$ in Example 8 above.) Be careful of overlapping notation, for example, when the particular object you have identified has precisely the same symbol as the one in the for-all statement you are specializing.

SUMMARY

This and the previous three chapters have provided various techniques for dealing with quantifiers that can appear in either A or B. As always, let the form of the statement guide you. When B contains the quantifier "there is," the construction method can be used to produce the desired object. The choose method is associated with the quantifier "for all" in the *backward* process, except when the statement B is supposed to be true for every *integer* starting from some initial one. In the latter case, induction is likely to be successful, provided that you can relate the statement for $(n + 1)$ to the one for n. Finally, if the quantifier "for all" appears in the *forward* process, you can use specialization. To do so, follow these steps:

1. Identify, in the for-all statement, the objects, the certain property, and the something that happens.
2. Look for one particular object to apply specialization to. (This object often arises from the backward process, especially when the choose method has been used.)
3. Verify that this particular object does have the certain property specified in the for-all statement.
4. Conclude, by writing a new statement in the *forward* process, that the something happens for this one particular object.

It is very common to confuse the choose method with the specialization method. Use the choose method when you encounter the key words "for all" in the *backward* process; use specialization when the key words "for all" arise in the *forward* process. Another way to say this is to use the choose method when you want to *show* that "for all objects with a certain property, something happens"; use specialization when you *know* that "for all objects with a certain property, something happens."

All of the statements thus far have contained only one quantifier. In the next chapter you will learn what to do when those statements contain more than one quantifier.

EXERCISES

Note: All proofs should contain an analysis of proof as well as a condensed version.

7.1. Explain the difference between the specialization and the choose methods.

7.2. Suppose you are working backward and you want to show that, for a particular object Y, something happens. Explain how specialization can be used to do so, that is, what form of statement should you look for in the forward process? How and when can you use that forward statement to reach the desired conclusion?

7.3. For each of the following statements and given objects, what properties must the given object satisfy in order to be able to apply specialization and, given that it does satisfy the properties, what can you conclude about the object?

a. Statement: $\forall$ integers $n \geq 5$, $2^n > n^2$.
   Given object: m.

b. Statement: For every real number x in the set S with $|x| < 5$, x is in the set T.
   Given object: y.

c. Statement: Any rectangle whose area is one-half the square of the length of a diagonal is a square.

Given object: the quadrilateral $QRST$.

d. Statement: For any angle t with $0 < t < \frac{\pi}{4}$, $\cos(t) > \sin(t)$.

Given object: Angle S of the triangle RST.

7.4. For each of the definitions in Exercise 5.1 on page 63, explain precisely how you would work forward from the associated for-all statements. For example, to work forward from the for-all statement in Exercise 5.1(b), one must (1) look for a specific element, say y, with which to apply specialization; (2) show that y is an element of the set S; and (3) conclude that $g(y) \geq f(y)$ as a new statement in the forward process.

7.5. Prove that if R is a subset of S and S is a subset of T, then R is a subset of T.

7.6. Write an analysis of proof that corresponds to the condensed proof given below. Indicate which proof techniques are being used and how they are applied. Fill in the details of any missing steps where appropriate.

Proposition. If R is subset of a set S of real numbers, and if f and g are functions for which $g \geq f$ on S [see the definition in Exercise 5.1(b)], then $g \geq f$ on R.

Proof. To show that $g \geq f$ on R, let $x \in R$. Since R is a subset of S, every element $r \in R$ is in S. In particular, $x \in R$, so $x \in S$. Also, since $g \geq f$ on S, it follows that for every element $s \in S$, $g(s) \geq f(s)$. In particular, $x \in S$, so $g(x) \geq f(x)$. ∎

7.7. Prove that if S and T are convex sets [see Exercise 5.1(d)], then S intersect T is a convex set. (Recall that S intersect $T = \{\text{elements } x : x \in S \text{ and } x \in T\}$.)

7.8. Explain what is wrong with the proof given below.

Proposition. If R is a nonempty subset of a set S of real numbers and R is convex [see Exercise 5.1(d)], then S is convex.

Proof. To show that S is a convex set, let $x,y \in S$, and t be a real number with $0 \le t \le 1$. Since R is a convex set, by definition, for any two elements u and v in R, and for any real number s with $0 \le s \le 1$, $su + (1 - s)v \in R$. In particular, for the specific elements x and y, and for the real number t, it follows that $tx + (1 - t)y \in R$. Since R is a subset of S, it follows that $tx + (1 - t)y$ is in S and so S is convex. ∎

7.9. Prove that if f is a convex function of one variable [see Exercise 5.1(e)], then for all real numbers $s \ge 0$, the function sf is convex [where the value of the function sf at any point x is $sf(x)$].

7.10. Prove that if f and g are convex functions of one variable [see Exercise 5.1(e)], then the function $f + g$ is a convex function [where the value of the function $f+g$ at any point x is $f(x) + g(x)$].

7.11. Prove that if f is a convex function of one variable and y is a given real number, then the set $C = \{\text{real numbers } x : f(x) \le y\}$ is convex [see Exercises 5.1(d and e)].

7.12. Prove that if a and b are real numbers, then the set $S = \{\text{real numbers } x : ax \le b\}$ is a convex set [see Exercise 5.1(d)].

EIGHT

QUANTIFIERS V: NESTED QUANTIFIERS

The statements in the previous four chapters contained only one quantifier—either "there is" or "for all." In this chapter you will learn what to do when statements contain more than one quantifier, in which case they are said to be **nested quantifiers**.

UNDERSTANDING STATEMENTS WITH NESTED QUANTIFIERS

The use of nested quantifiers is illustrated in the following statement containing both "for all" and "there is":

S1: For all real numbers x with $0 \leq x \leq 1$, there is a real number y with $-1 \leq y \leq 1$ such that $x + y^2 = 1$.

When reading, writing, or processing such statements, *always work from left to right.* For the statement $S1$ above, the first quantifier encountered from the left is "for all." For that quantifier, you should identify the following components:

Object: real number x.
Certain property: $0 \leq x \leq 1$.
Something happens: there is a real number y with $-1 \leq y \leq 1$ such that $x + y^2 = 1$.

The something that happens above contains the (nested) quantifier "there is," for which you can then identify its three components:

Object: real number y.
Certain property: $-1 \leq y \leq 1$.
Something happens: $x + y^2 = 1$.

As another example of nested quantifiers, suppose that T is a set of real numbers and consider the statement:

S2: There is a real number $M > 0$ such that for all elements x in the set T, $x < M$.

In $S2$, the quantifier "there is" is the first one encountered when reading from the left. Associated with that quantifier are its three components:

Object: real number M.
Certain property: $M > 0$.
Something happens: for all elements x in the set T, $x < M$.

The something that happens above contains the (nested) quantifier "for all," and you can identify its three components:

Object: element x.
Certain property: x in the set T.
Something happens: $x < M$.

It is important to realize that *the order in which the quantifiers appear can be critical to the meaning of the statement.* For example, compare the statement $S2$ above with the following statement:

S3: For all real numbers $M > 0$, there is an element x in the set T such that $x < M$.

$S3$ states that for each positive real number M, you can find a (possibly different) element x in T for which $x < M$. (Note that the element x may depend on the value of M, that is, if you change the value of M, you may have to change the value of x).

In contrast, $S2$ says that there is a positive real number M such that, no matter which element x you choose in T, $x < M$. It should be clear that $S2$ and $S3$ are not the same.

Statements can have any number of quantifiers, as illustrated in the following example containing three nested quantifiers (in which f is a function of one variable):

S4: For every real number $\varepsilon > 0$, there is a real number $\delta > 0$ such that for all real numbers x and y with $|x - y| < \delta$, $|f(x) - f(y)| < \varepsilon$.

Applying the principle of working from left to right, associated with the first quantifier "for all" in $S4$ you should identify:

Object: real number ε.
Certain property: $\varepsilon > 0$.
Something happens: there is a real number $\delta > 0$ such that for all real numbers x and y with $|x - y| < \delta$, $|f(x) - f(y)| < \varepsilon$.

The something that happens above contains the nested quantifiers "there is" and "for all." Working from left to right once again, identify the three components associated with the quantifier "there is":

Object: real number δ.
Certain property: $\delta > 0$.
Something happens: for all real numbers x and y with $|x - y| < \delta$, $|f(x) - f(y)| < \varepsilon$.

The three components of the last nested quantifier "for all" in the something that happens above are:

Objects: real numbers x and y.
Certain property: $|x - y| < \delta$.
Something happens: $|f(x) - f(y)| < \varepsilon$.

USING PROOF TECHNIQUES WITH NESTED QUANTIFIERS

When a statement in the forward or backward process contains nested quantifiers, apply appropriate techniques (construction, choose, induction, or specialization) *based on the order of the quantifiers from left to right in the statement.* To illustrate, suppose you want to show that the statement $S2$ above is true, that is, you want to show that

B: there is a real number $M > 0$ such that for all elements x in the set T, $x < M$.

Since the statement B is in the backward process, and since the first quantifier encountered from the left is "there is," the construction method is the correct proof technique to use first. In this case, you would turn to the forward process in an attempt to construct a real number M. Suppose you have done so. According to the construction method, you must show that the value of M you constructed satisfies the certain property and the something that happens, that is, you must show that

B1: $M > 0$

and

B2: for all elements x in the set T, $x < M$.

When you try to show that $B2$ is true, you should apply the choose method because of the appearance of the quantifier "for all" in the backward process. In this case, you would choose

A1: an element x in T,

for which it must be shown that

B3: $x < M$.

Applying proof techniques to nested quantifiers is illustrated again in the following complete example.

Definition 17. A function of one variable is **onto** if and only if for every real number y, there is a real number x such that $f(x) = y$.

EXAMPLE 9.

Proposition. If m and b are real numbers with $m \neq 0$, then the function $f(x) = mx + b$ is onto.

Analysis of Proof. The forward–backward method is used to begin the proof since the hypothesis A and the conclusion B do not contain key words (such as "for all" or "there is"). Working backward, you are led to the key question: "How can I show that a function [namely, $f(x) = mx + b$] is onto?" Applying Definition 17 to the specific function $f(x) = mx + b$, you must show that:

B1: for every real number y, there is a real number x such that $mx + b = y$.

The statement $B1$ contains nested quantifiers. In deciding which proof technique to apply next, observe that the quantifier "for all" is the *first* one encountered from left to right. Thus, the *choose* method should be used (since the key words "for all" appear in the backward process). Accordingly, you should choose

A1: a real number y,

for which it must be shown that

B2: there is a real number x such that $mx + b = y$.

Recognizing the key words "there is" in $B2$, you should now proceed with the construction method. To that end, turn to the forward process in an attempt to construct the desired real number x.

Looking at the fact that you want $mx + b = y$ and observing that $m \neq 0$ by the hypothesis, you might produce this statement:

A2: Construct the real number $x = \dfrac{(y - b)}{m}$.

Recall that, with the construction method, you must show that the object you constructed satisfies the certain property and the something that happens (in $B2$). Thus you must show that

B3: $mx + b = y$.

But, from $A2$,

$$\begin{aligned} \textbf{A3: } mx + b &= m\left[\frac{(y-b)}{m}\right] + b \\ &= (y-b) + b \\ &= y, \end{aligned}$$

and so the proof is complete.

In the condensed proof that follows, observe that the names of the proof techniques are omitted altogether. Note also that several steps of the proof given above have been combined into a single statement.

Proof of Example 9. To show that f is onto, let y be a real number. (The word "let" here indicates that the choose method has been used.) Since, by hypothesis, $m \neq 0$, let $x = (y - b)/m$. (Here, the word "let" indicates that the construction method has been used.) It is easy to see that $f(x) = mx + b = y$, and so the proof is complete. ∎

SUMMARY

When working with statements containing nested quantifiers, follow these steps:

1. For each quantifier encountered from left to right, identify the object, the certain property, and the something that happens.
2. Apply appropriate proof techniques such as construction, choose, induction, and/or specialization based on the order of the quantifiers as they appear from left to right.

All the material thus far has been organized around the forward–backward method. Now it is time to see some other techniques for showing that "A implies B."

EXERCISES

Note: All proofs should contain an analysis of proof as well as a condensed version.

8.1. Identify the objects, the certain property, and the something that happens for each of the quantifiers as they appear from left to right in each of the following definitions.

a. The function f of one variable is **bounded above** if and only if there is a real number y such that for every real number x, $f(x) \leq y$.

b. The real number u is a **least upper bound** for a set S of real numbers if and only if u is an upper bound for S and $\forall$ real numbers $\varepsilon > 0$, $\exists x \in S \ni x > u - \varepsilon$.

c. The function f of one variable is **continuous at the point x** if and only if for every real number $\varepsilon > 0$, there is a real number $\delta > 0$ such that, for all real numbers y with $|x - y| < \delta$, $|f(x) - f(y)| < \varepsilon$.

d. Suppose that $x_1, x_2, \ldots$ are real numbers. The sequence $x_1, x_2, \ldots$ **converges** to the real number x if and only if $\forall$ real numbers $\varepsilon > 0$, $\exists$ an integer $k' \ni \forall$ integers k with $k > k'$, $|x_k - x| < \varepsilon$.

8.2. Consider the two statements:

S1: For every object X with a certain property P, there is an object Y with a certain property Q such that something happens.

S2: There is an object X with a certain property P such that for every object Y with a certain property Q, something happens.

a. Create one specific mathematical statement for which $S1$ is true and $S2$ is false.

b. Create one specific mathematical statement for which $S1$ is false and $S2$ is true.

c. Create one specific mathematical statement for which $S1$ and $S2$ are both true.

8.3. Are each of the following pairs of statements the same, that is, are they both true or both false? Why or why not? Explain.

a. **S1:** For all real numbers x with $0 \leq x \leq 1$, for all real numbers y with $0 \leq y \leq 2$, $2x^2 + y^2 \leq 6$.

S2: For all real numbers y with $0 \leq y \leq 2$, for all real numbers x with $0 \leq x \leq 1$, $2x^2 + y^2 \leq 6$.

b. **S1:** For all real numbers x with $0 \leq x \leq 1$, for all real numbers y with $0 \leq y \leq 2x$, $2x^2 + y^2 \leq 6$.

S2: For all real numbers y with $0 \leq y \leq 1$, for all real numbers x with $0 \leq x \leq 2y$, $2x^2 + y^2 \leq 6$.

c. Based on your answer to parts (a) and (b), when is it true that the statement "for all objects X with a certain property, say P, and for all objects Y with a certain property, say Q, something happens," is the same as the statement, "for all objects Y with the property Q, and for all objects X with the property P, something happens"?

8.4. Are each of the following pairs of statements the same, that is, are the sets of values for the real numbers x and y that make both the statements true the same? Why or why not? Explain.

a. **S1:** There is a real number $x > 2$ such that, there is a real number $y > 1$ such that $2x^2 + y^2 > 10$.

S2: There is a real number $y > 1$ such that, there is a real number $x > 2$ such that $2x^2 + y^2 > 10$.

b. **S1:** There is a real number $x > 2$ such that, there is a real number $y > 2x$ such that $2x^2 + y^2 > 24$.

S2: There is a real number $y > 2$ such that, there is a real number $x > 2y$ such that $2x^2 + y^2 > 24$.

c. Based on your answers to parts (a) and (b), when is it

true that the statement "there is an object X with a certain property, say P, such that, there is an object Y with a certain property, say Q, such that something happens," is the same as the statement, "there is an object Y with the property Q such that, there is an object X with the property P such that something happens"?

8.5. For each of the following statements *in the backward process*, indicate which proof techniques you would use (choose and/or construction) and in which order. Also, explain how the techniques would be applied to the particular problem, that is, what would you construct, what would you choose, etc.

a. There is a real number $M > 0$ such that, for all elements x in the set T of real numbers, $|x| \leq M$.

b. For all real numbers $M > 0$, there is an element x in the set T of real numbers such that $|x| > M$.

c. $\forall$ real numbers $\varepsilon > 0$, $\exists$ a real number $\delta > 0 \ni \forall$ real numbers x and y with $|x - y| < \delta$, $|f(x) - f(y)| < \varepsilon$ (where f is a function of one variable).

8.6. Explain how to work *forward* from each of the following statements. How would the proof techniques be applied?

a. For all objects X with a certain property P, there is an object Y with a certain property Q such that something happens.

b. There is an object X with a certain property P such that, for all objects Y with a certain property Q, something happens.

8.7. Prove that for every real number $x > 2$, there is a real number $y < 0$ such that $x = 2y/(1 + y)$.

8.8. Prove that the function $f(x) = -x^2 + 2x$ is bounded above [see Exercise 8.1(a)].

8.9. Prove that 1 is a least upper bound of the set

$S = \{1 - \frac{1}{2}, 1 - \frac{1}{3}, 1 - \frac{1}{4}, \ldots\}$ [See Exercises 8.1(b) and 5.1(c)].
(Hint: The set S can be written as
{real numbers x : there is an integer $n \geq 2$ such that $x = 1 - \frac{1}{n}$}.)

NINE

THE CONTRADICTION METHOD

With all the proof techniques you have learned so far, you may well find yourself unable to complete a proof for one reason or another. This chapter presents a new technique that can often be used when the conclusion of your problem contains appropriate key words.

WHY THE NEED FOR ANOTHER PROOF TECHNIQUE?

As powerful as the forward–backward method is, it may not always lead to a successful proof, as shown in the next example.

EXAMPLE 10.

Proposition. If n is an integer and n^2 is even, then n is even.

Analysis of proof. The forward–backward method gives rise to the key question "How can I show that an integer (namely, n) is even?" One answer is to show that

B1: there is an integer k such that $n = 2k$.

The appearance of the quantifier "there is" in the backward process suggests proceeding with the construction method, and so the forward process will be used in an attempt to produce the desired integer k.

Working forward from the hypothesis that n^2 is even, you can state that

A1: there is an integer, say m, such that $n^2 = 2m$.

Since the objective is to produce an integer k for which $n = 2k$, it is natural to take the positive square root of both sides of the equality in $A1$ to obtain

A2: $n = \sqrt{2m}$,

but how can you rewrite $\sqrt{2m}$ to look like $2k$? It would seem that the forward–backward method has failed!

***Proof of Example 10*.** The technique you are about to learn will lead to a simple proof of this problem, and it is left as an exercise. ∎

Fortunately, there are several other techniques that you might want to try before you give up. In this chapter, the **contradiction method** is described together with an indication of how and when it should be used.

HOW AND WHEN TO USE THE CONTRADICTION METHOD

With the contradiction method, you begin by assuming that A is true, just as you would do in the forward–backward method. However, to reach the desired conclusion that B is true, you proceed by asking yourself a simple question: "Why can't B be false?" After all, if B is supposed to be true, then *there must be some reason why B cannot be false.* The objective of the contradiction method is to discover that reason.

In other words, the idea of a proof by contradiction is to assume that A is true and B is false, and see why this cannot happen. So what does it mean to "see why this cannot happen"? Suppose, for example, that as a result of assuming that A is true and B is false (hereafter written as NOT B), you were somehow able to reach the conclusion that $1 = 0$!?! Would that not convince you that it is impossible for A to be true and B to be false simultaneously? Thus, in a proof by contradiction, you assume that A is true and that NOT B is true. You must use this information to reach a contradiction to something that you absolutely know to be true.

Another way of viewing the contradiction method is to recall from Table 1 on page 6 that the statement "A implies B" is true in all cases *except* when A is true and B is false. In a proof by contradiction, you rule out this one unfavorable case by assuming that it actually does happen, and then reaching a contradiction.

At this point, several natural questions arise:

1. What contradiction should you be looking for?
2. Exactly how do you use the assumption that A is true and B is false to reach the contradiction?
3. Why and when should you use this approach instead of the forward–backward method?

The first question is by far the hardest to answer because there are no specific guidelines. Each problem gives rise to its own contradiction; it usually takes creativity, insight, persistence, and luck to produce a contradiction.

As to the second question, the most common approach to finding a contradiction is to work forward from the assumptions that A and NOT B are true, as will be illustrated in a moment.

The discussion above also indicates why you might wish to use contradiction instead of the forward–backward method. With the forward–backward method you assume only that A is true, while in the contradiction method, you can assume that both A and NOT B are true. Thus, you get *two* statements from which to reason forward instead of just one (see Figure 9.1). On the other hand, you have no definite knowledge of where the contradiction will arise.

As a general rule, use contradiction when the statement NOT

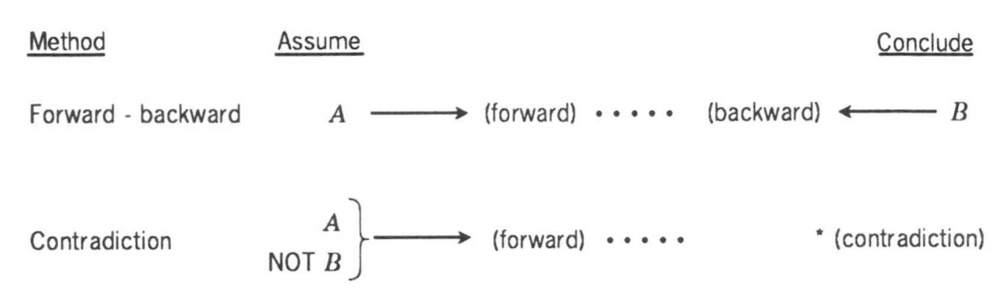

Figure 9.1 Forward–Backward vs. Contradiction

B gives you some "useful" information. There are at least two recognizable instances when this will happen. Recall the statement B associated with Example 10: "n is an even integer." Obviously an integer can only be odd or even. When you assume that B is not true (i.e., that n is *not* an even integer), then it *must* be the case that n is an odd integer. Here, the statement NOT B has given you some useful information. In general, when the statement B is one of two possible alternatives (as in Example 10), the contradiction method is likely to be effective because, by assuming NOT B, you will know that the other case must happen, and that should help you to reach a contradiction.

A second instance when the contradiction method is likely to be successful is when the statement B contains the word "no" or "not," as is shown in the next example.

EXAMPLE 11.

Proposition. If r is a real number such that $r^2 = 2$, then r is irrational.

Analysis of proof. It is important to note that the conclusion of Example 11 can be rewritten so as to read "r is not rational." In this form, the appearance of the word "not" now suggests using the contradiction method, whereby you can assume that A and NOT B are both true. In this case that means you can assume that

A: $r^2 = 2$, and

A1: r *is* a rational number (NOT B).

A contradiction must now be reached using this information.

Working forward from $A1$ by using Definition 7 on page 30 for a rational number,

A2: there are integers p and q with $q \neq 0$ such that $r = \frac{p}{q}$.

There is still the unanswered question of where the contradiction arises, and this takes a lot of creativity. A crucial observation here will really help. It is possible to assume that

A3: p and q have no common divisor
(i.e., there is no integer that divides both p and q).

The reason for this is that if p and q did have a common divisor, you could divide this integer out of both p and q.

Now a contradiction can be reached by showing that 2 is a common divisor of p and q! This will be done by working forward to show that p and q are even, and hence 2 divides them both.

Working forward by squaring both sides of the equality in $A2$, it follows that

A4: $r^2 = \frac{p^2}{q^2}$.

But from A you also know that $r^2 = 2$, so

A5: $2 = \frac{p^2}{q^2}$.

The rest of the forward process is mostly rewriting $A5$ via algebraic manipulations to reach the desired contradiction that both p and q are even integers. Specifically, on multiplying both sides of $A5$ by q^2 you obtain

A6: $2q^2 = p^2$.

Looking at $A6$, you can certainly say that, no matter what kind of integer q is, $2q^2$ is even, and since $p^2 = 2q^2$, it follows that

A7: p^2 is even.

Continuing with the thc forward process, what useful information can be derived from the fact that p^2 is even? Well, the only way for the integer p times itself to be even is for p to be even, and so you have the statement:

A8: p is even.

(Recall here that a proof is supposed to be a convincing argument, and this last sentence might not have convinced you that p is even. In this case it would be necessary to provide more details. As always, a proof should be written with the audience in mind. If you need further convincing that p is even, look at Example 10.)

It is still necessary to show that q is even too in order to reach a contradiction. Continuing with the forward process from $A8$, since p is even you know that

A9: there is an integer k such that $p = 2k$.

Replacing p with $2k$ in $A6$ ($2q^2 = p^2$), it follows that

A10: $2q^2 = (2k)^2 = 4k^2$,

which, on dividing both sides by 2, says that

A11: $q^2 = 2k^2$.

Once again, $2k^2$ is even no matter what type of integer k is, and since $q^2 = 2k^2$, it follows that

A12: q^2 is even.

Finally, note that the only way for an integer q times itself to be even is for q to be even (see Example 10), and so,

A13: q is even.

So, both p and q are even (see $A8$ and $A13$), and this contradicts $A3$, thus completing the proof.

***Proof of Example 11*.** Assume, to the contrary, that r is a rational number of the form $\frac{p}{q}$ (where p and q are integers with $q \neq 0$) and that $r^2 = 2$. Furthermore, it can be assumed that p and q have no common divisor for, if they did, this number could be canceled from both the numerator p and the denominator q. Since $r^2 = 2$ and $r = \frac{p}{q}$, it follows that $2 = \frac{p^2}{q^2}$, or equivalently, $2q^2 = p^2$. Noting that $2q^2$ is even, p^2, and hence p, must be even. Thus, there is an integer k such that $p = 2k$. On substituting this value for p one obtains $2q^2 = p^2 = (2k)^2 = 4k^2$, or equivalently, $q^2 = 2k^2$. From this it then follows that q^2, and hence q, must be even. Thus it has been shown that both p and q are even and have the common divisor 2. This contradiction establishes the claim. ∎

This proof was discovered in ancient times by a follower of Pythagoras, and it epitomizes the use of contradiction. Try to prove the statement by some other method!

ADDITIONAL USES FOR THE CONTRADICTION METHOD

There are several other valuable uses for the contradiction method. Recall that, when the statement B contains the quantifier "there is," the construction method is recommended in spite of the difficulty of actually having to produce the desired object. The contradiction method opens up a whole new approach. Instead of trying to show that there is an object with the certain property such that the something happens, why not proceed from the assumption that there is *no* such object? Now your job is to use this information to reach some kind of contradiction. How and where the contradiction arises is not at all clear, but it may be a lot easier than producing or constructing the object. Consider the following example.

Suppose you wish to show that at a party of 367 people, there are at least two people whose birthday falls on the same day of the year. If the construction method is used, then you would actually have to go to the party and find two such people. To save you the time and trouble, the contradiction method can be used. In so doing, you can assume that no two people's birthday fall on the same day of the year, or equivalently, that everyone's birthday falls on a different day of the year.

To reach a contradiction, assign numbers to the people in such a way that the person with the earliest birthday of the year receives the number 1, the person with the next earliest birthday receives the number 2, and so on. Recall that each person's birthday is assumed to fall on a different day. Thus, the birthday of the person whose number is 2 must occur at least one day later than the person whose number is 1, and so on. Consequently, the birthday of the person whose number is 367 must occur at least 366 days after the person whose number is 1. But a year has at most 366 days, and so this is impossible, that is, a contradiction has been established.

This example illustrates a subtle but very significant difference between a proof using the construction method and one that uses contradiction. If the construction method is successful, then you will have produced the desired object, or at least indicated how it might be produced, perhaps with the aid of a computer. On the other hand, if you establish the same result by contradiction, then you will know that the object exists, but will have no way of physically constructing it. For this reason, it is often the case that proofs done by contradiction are quite a bit shorter and easier than those done by construction because you do not have to create the desired object. You only have to show that its *nonexistence* is impossible! This difference has led to some great philosophical debates in mathematics. Moreover, an active area of current research consists of finding constructive proofs where previously only proofs by contradiction were known.

SUMMARY

The contradiction method can be a very useful technique when the statement B contains the key word "no" or "not." To use this method, follow these steps:

1. Assume that A is true.
2. Assume that B is not true (i.e., assume NOT B is true).
3. Work forward from A and NOT B to reach a contradiction.

One of the disadvantages of this method is that you do not know exactly what the contradiction is going to be. The next chapter describes another proof technique in which you attempt to reach a very specific contradiction. As such, you will have a "guiding light" since you will know what contradiction you are looking for.

EXERCISES

Note: All proofs should contain an analysis of proof as well as a condensed version.

9.1. When applying the contradiction method to the following propositions, what should you assume?

a. If l, m, and n are three consecutive integers, then 24 does not divide $l^2 + m^2 + n^2 + 1$.

b. If the matrix M is not singular, then the rows of M are not linearly dependent.

c. If f and g are two functions such that (1) $g \geq f$ and (2) f is unbounded above, then g is unbounded above.

9.2. a. What statement(s) would you work backward from when using the contradiction method to show that "A implies B"?

b. A mathematics student who applied the contradiction method to show that "A implies B" said, at the end of the proof, "...and since I have been able to show that A is not true, the proof is complete." Do you agree with the student? Why or why not? Explain.

9.3. Reword each of the following statements so that the word "not" appears explicitly.

a. There are an infinite number of primes.

b. The set S of real numbers is unbounded.

c. The only positive integers that divide the positive integer p are 1 and p.

d. The lines ℓ and ℓ' in a plane P are parallel.

e. The real number x is < 5.

9.4. Reword each of the following statements so that the word "not" does not appear.

a. There are not a finite number of primes.

b. The real number x is not < 5.

c. The lines ℓ and ℓ' in a plane P are not parallel.

9.5. When trying to prove each of the following statements, which proof techniques would you use, and in which order? Specifically, state what you would assume and what you would try to conclude. (Throughout, S and T are sets of real numbers, and all the variables refer to real numbers.)

a. $\exists s \in S \ni s \in T$.

b. $\forall s$ in S, $\nexists t$ in T such that $s > t$.

c. $\nexists M > 0$ such that, $\forall x$ in S, $|x| < M$.

9.6. Prove, by contradiction, that if ℓ_1 and ℓ_2 are two lines in a plane that are both perpendicular to a third line ℓ in the plane, then ℓ_1 and ℓ_2 are parallel. (Hint: Recall that the sum of the degrees of the angles in a triangle is 180.)

9.7. Prove, by contradiction, that if n is an integer and n^2 is even, then n is even.

9.8. Prove, by contradiction, that if n is an integer and n^2 is odd, then n is odd.

9.9. Prove, by contradiction, that no chord of a circle is longer than a diameter.

9.10. Prove, by contradiction, that if $n - 1$, n, and $n + 1$ are consecutive positive integers, then the cube of the largest cannot be equal to the sum of the cubes of the other two.

9.11. Write an analysis of proof that corresponds to the condensed proof given below. Indicate which proof techniques are being used and how they are applied. Fill in the details of any missing steps where appropriate.

Proposition. If a, b, c, x, y, and z are real numbers with $b \neq 0$ such that (1) $az - 2by + cx = 0$ and (2) $ac - b^2 > 0$, then it must be that $xz - y^2 \leq 0$.

Proof. Assume that $xz - y^2 > 0$. From this and (2) it follows that $(ac)(xz) > b^2y^2$. Rewriting (1) and squaring both sides one obtains $(az + cx)^2 = 4b^2y^2 < 4(ac)(xz)$. Rewriting, one has that $(az - cx)^2 < 0$, which cannot happen. ∎

9.12. Write an analysis of proof that corresponds to the condensed proof given below. Indicate which proof techniques are being used and how they are applied. Fill in the details of any missing steps where appropriate.

Proposition. The polynomial $x^4 + 2x^2 + 2x + 2$ cannot be expressed as the product of the two polynomials $x^2 + ax + b$ and $x^2 + cx + d$ in which a, b, c, and d are integers.

Proof. Suppose that

$$\begin{aligned} x^4 + 2x^2 + 2x + 2 &= (x^2 + ax + b)(x^2 + cx + d) \\ &= x^4 + (a + c)x^3 + (b + ac + d)x^2 + \\ &\qquad (bc + ad)x + bd. \end{aligned}$$

It would then follow that the integers a, b, c, and d must satisfy

1. $a + c = 0$.
2. $b + ac + d = 2$.
3. $bc + ad = 2$.
4. $bd = 2$.

The only way (4) can happen is if one of the factors b or d is odd (± 1) and the other is even (± 2). Suppose that

b is in fact odd and d is even. From (3), it would then follow that c is even, but then the left side of (2) would be odd, which is impossible. A similar contradiction can be reached if b is even and d is odd. ∎

9.13. Prove, by contradiction, that at a party of n (≥ 2) people, there are at least two people who have the same number of friends at the party.

9.14. Prove, by contradiction, that there are at least two people in the world who have exactly the same number of hairs on their head. (Hint: You can assume than no one has more than one billion hairs.)

9.15. Prove, by contradiction, that there are an infinite number of primes. (Hint: Assume that n is the largest prime. Then consider any prime number p that divides $n! + 1$. How is p related to n?)

9.16. Prove, by contradiction, that if m and n are odd integers, then the equation $x^2 + 2mx + 2n = 0$ has no rational root. To do so, prove each of the following two lemmas by contradiction; then use the approach in part (c).

a. If m and n are odd integers, then the above equation has no root that is an odd integer.

b. If m and n are odd integers, then the above equation has no root that is an even integer.

c. Now assume that the equation has a rational root. Rewrite the equation in such a way that a *new* equation of the form $y^2 + 2m'y + 2n'$ (where m' and n' are odd integers) now has an integer solution.

TEN

THE CONTRAPOSITIVE METHOD

The previous chapter described the contradiction method in which you work forward from the two statements A and NOT B to reach some kind of contradiction. The difficulty with this method is that you do not know what the contradiction is going to be. As will be seen in this chapter, the **contrapositive method** has the advantage of directing you toward one specific contradiction.

HOW AND WHEN TO USE THE CONTRAPOSITIVE METHOD

The contrapositive method is similar to contradiction in that you begin by assuming that A and NOT B are true. Unlike contradiction, however, you do not work forward from both A and NOT B. Instead, you work forward only from NOT B. Your objective is to reach the contradiction that A is false (hereafter written NOT A). Can you ask for a better contradiction than that? How can A be true and false at the same time? To repeat, in the contrapositive method you assume that A and NOT B are true; you then work forward from the statement NOT B to reach the contradiction that A is false.

The contrapositive method can be thought of as a "passive"

form of contradiction in the sense that the assumption that A is true passively provides the contradiction. In the contradiction method, the assumption that A is true is used actively to reach a contradiction. The next example demonstrates the contrapositive method.

Definition 18. A function f of one variable is **one-to-one** if and only if for all real numbers x and y with $x \neq y$, $f(x) \neq f(y)$.

EXAMPLE 12.

Proposition. If m and b are real numbers with $m \neq 0$, then the function $f(x) = mx + b$ is one-to-one.

Analysis of proof. The forward–backward method is used to begin the proof because the hypothesis A and the conclusion B do not contain any key words (such as "for all," "there is," "no," "not," etc.). The key question associated with B is: "How can I show that a function is one-to-one?" Applying Definition 18 to the particular function $f(x) = mx + b$ means you must now show that

B1: for all real numbers x and y with $x \neq y$, $mx + b \neq my + b$.

This new statement contains both the key words "for all" and "not" (in "not equal"). When more than one group of key words is present in a particular statement, apply appropriate proof techniques based on the occurrence of the key words as they appear from *left to right* (just as you learned to do with nested quantifiers in Chapter 8). Since the first key words from the left in the (backward) statement $B1$ are "for all," the choose method should be used next. Accordingly, you should choose

A1: real numbers x and y with $x \neq y$

for which it must be shown that

B2: $mx + b \neq my + b$.

Recognizing the key word "not" in $B2$, it is appropriate to proceed with either the contradiction or contrapositive method. In this case, the contrapositive method will be used. Thus, you should work forward from NOT $B2$:

A2: $mx + b = my + b$ (NOT $B2$)

and backward from NOT $A1$ (which is the last statement in the forward process):

B3: $x = y$ (NOT $A1$).

The remainder of the proof is simple algebra applied to $A2$. Specifically, on subtracting b from both sides of $A2$ you obtain:

A3: $mx = my$.

Finally, since $m \neq 0$ by the hypothesis A, dividing both sides of $A3$ by m yields precisely $B3$, and so the proof is complete.

In the condensed proof that follows, note that no mention is made of the choose or contrapositive methods.

Proof of Example 12. Let x and y be real numbers for which $mx + b = my + b$. It will be shown that $x = y$. But this follows by subtracting b from both sides and then dividing by m (noting that $m \neq 0$). ∎

COMPARING THE CONTRAPOSITIVE METHOD

As mentioned above, the contrapositive method is a type of proof by contradiction. Each of these methods has its advantages and disadvantages, as illustrated in Figure 10.1. The disadvantage of the contrapositive method is that you work forward from only one statement (namely, NOT B) instead of two. On the other hand, the advantage is that you know precisely what you are looking for (namely, NOT A). Thus you can apply the backward process to the statement NOT A. The option of working backward is not available in the contradiction method because you do not know what contradiction you are looking for.

It is also quite interesting to compare the contrapositive and forward–backward methods. In the forward–backward method

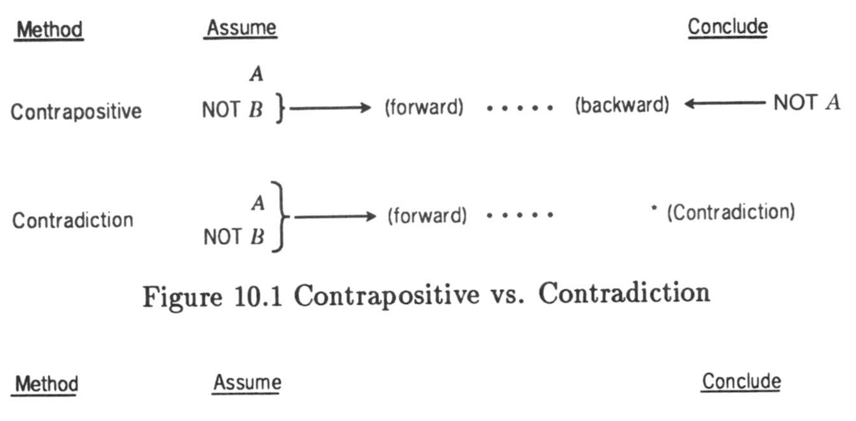

Figure 10.1 Contrapositive vs. Contradiction

Method	Assume			Conclude
Forward - backward	A ⟶	(forward) ·····	(backward) ⟵	B
Contrapositive	A, NOT B ⟶	(forward) ·····	(backward) ⟵	NOT A

Figure 10.2 Forward-Backward vs. Contrapositive

you work forward from A and backward from B; in the contrapositive method, you work forward from NOT B and backward from NOT A (see Figure 10.2).

From Figure 10.2, it is not hard to see why the contrapositive method might be better than the forward–backward method. Perhaps you can obtain more useful information by working forward from NOT B rather than from A. It might also be easier to work backward from the statement NOT A rather than from B, as would be done in the forward–backward method.

The forward–backward method arose from considering what happens to the truth of "A implies B" when A is true and when A is false (recall Table 1 on page 6). The contrapositive method arises from similar considerations regarding B. Specifically, if B is true then, according to Table 1, the statement "A implies B" is true. Hence, there is no need to consider the case when B is true. So suppose B is false. In order to ensure that "A implies B" is true, according to Table 1, you would have to show that A

is false. Thus, the contrapositive method has you assume that B is false and try to conclude that A is false.

The statement "A implies B" is logically equivalent to "NOT B implies NOT A" (see Table 4 on page 39). The contrapositive method can therefore be thought of as the forward–backward method applied to the statement "NOT B implies NOT A." Most condensed proofs that use the contrapositive method make little or no reference to the contradiction method.

In general, it is difficult to know whether the forward–backward, contradiction, or contrapositive method will be more effective for a given problem without trying each one. However, there is one instance that often indicates that the contradiction or contrapositive method should be chosen, or at least considered seriously. This occurs when the statement B contains the word "no" or "not" in it, for then you will usually find that the statement NOT B has some useful information.

In the contradiction method, you work forward from the two statements A and NOT B to obtain a contradiction. In the contrapositive method you also reach a contradiction, but you do so by working forward from NOT B to reach the conclusion NOT A; you should also work backward from NOT A. A comparison of the forward–backward, contradiction, and contrapositive methods is given in Figure 10.3.

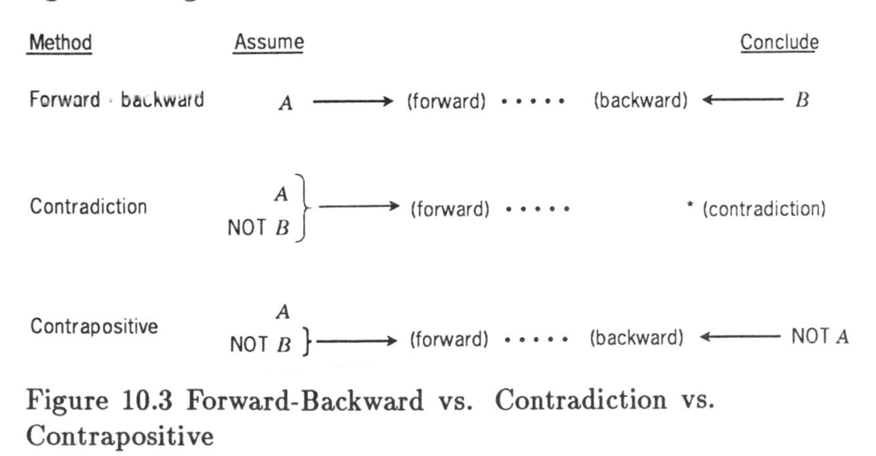

Figure 10.3 Forward-Backward vs. Contradiction vs. Contrapositive

SUMMARY

The contrapositive method, being a type of proof by contradiction, should be used when the last statement in the backward process contains the key word "no" or "not." With the contrapositive method, you work toward a specific, known contradiction by

1. Assuming that A and NOT B are true.
2. Working forward from NOT B in an attempt to obtain NOT A.
3. Working backward from NOT A in an attempt to reach NOT B.

Both the contrapositive and contradiction methods require that you be able to write down the NOT of a statement. The next chapter shows you how to do so when the statements contain quantifiers.

EXERCISES

Note: All proofs should contain an analysis of proof as well as a condensed version.

10.1. If the contrapositive method is used to prove the following propositions, then what statement(s) will you work forward from and what statement(s) will you work backward from?

a. If n is an integer for which n^2 is even, then n is even.
b. Suppose that S is a subset of the set T of real numbers. If S is not bounded then T is not bounded.
c. If p and q are positive real numbers such that $\sqrt{pq}$ is not equal to $(p+q)/2$, then p is not equal to q.
d. If the matrix M is not singular, then the rows of M are not linearly dependent.

10.2. Bob said, "If I do not get at least eight hours of sleep, I cannot think straight the next day," to which Mary replied, "Well, since you are not thinking straight today, I guess you got less than eight hours of sleep." Was Mary's statement correct? Why or why not? Explain.

10.3. In a proof by the contrapositive method of the proposition "If r is a real number with $r > 1$, then there is no real number t with $0 < t < \frac{\pi}{4}$ such that $\sin(t) = r\cos(t)$," which of the following is a result of the forward process?

a. $r - 1 \leq 0$.

b. $\sin^2(t) = r^2(1 - \sin^2(t))$.

c. $1 - r < 0$.

d. $\tan(t) = \frac{1}{r}$.

10.4. Suppose that f is a function of one variable, S is a set of real numbers, and that the contrapositive method is used to prove the proposition: "If no element x in S satisfies the property that $f(x) = 0$, then there is no real number $M > 0$ such that for all elements x in S, $|x| < M$." Which of the following proof techniques will subsequently be used: construction, choose, specialization? Explain.

10.5. If the contrapositive method is used to prove the proposition "If the derivative of the function f at the point x is not equal to 0, then x is not a local maximizer of f," then which of the following is the correct key question? What is wrong with the other choices?

a. How can I show that the point x is a local maximizer of the function f?

b. How can I show that the derivative of the function f at the point x is 0?

c. How can I show that a point is a local maximizer of a function?

d. How can I show that the derivative of a function at a point is 0?

10.6. Prove, by the contrapositive method, that if p and q are positive real numbers with the property that $\sqrt{pq}$ is not equal to $(p+q)/2$, then p is not equal to q.

10.7. Prove, by the contrapositive method, that if c is an odd integer then the equation $n^2 + n - c = 0$ has no integer solution for n.

10.8. Prove, by the contrapositive method, that if n is an integer greater than 2, then there is no integer m with $n + m = nm$ such that $n|m$ (see Definition 1 on page 30).

10.9. Prove, by the contrapositive method, that if no angle of a quadrilateral $RSTU$ is obtuse, then the quadrilateral $RSTU$ is a rectangle.

10.10. Write an analysis of proof that corresponds to the condensed proof given below. Indicate which proof techniques are being used and how they are applied. Fill in the details of any missing steps where appropriate.

Proposition. Assume that a and b are integers with $a \neq 0$. If a does not divide b, then $ax^2 + bx + (b - a)$ has no positive integer root (see Definition 1 on page 30).

Proof. Suppose that x is a positive integer satisfying $ax^2+bx+(b-a) = 0$. Then $x = [-b\pm(b-2a)]/(2a)$. Since $x > 0$, it must be that $x = 1 - \frac{b}{a}$. But then $b = (1 - x)a$, and so $a|b$. Thus the proof is complete. ∎

10.11. Write an analysis of proof that corresponds to the condensed proof given below. Indicate which proof techniques are being used and how they are applied. Fill in the details of any missing steps where appropriate.

Proposition. Suppose that p is a positive integer. If there is no integer m with $1 < m \leq \sqrt{p}$ such that $m|p$, then p is prime (see Definitions 1 and 2 on page 30).

Proof. Assume that p is not prime, that is, that there is an integer n with $1 < n < p$ such that $n|p$. In the event that $n \leq \sqrt{p}$, then n is the desired integer m. Otherwise,

$n > \sqrt{p}$. Since $n|p$, there is an integer k such that $p = nk$. It then follows that k is the desired value for m. This is because $k > 1$, for otherwise, $p \leq n$. Also, $k \leq \sqrt{p}$, for otherwise, $nk > \sqrt{p}\sqrt{p} = p$. ∎

ELEVEN

NOTS OF NOTS LEAD TO KNOTS

As you saw in the previous chapter, the contrapositive method is a valuable proof technique. To use it, however, you must be able to write down the statement NOT B so that you can work forward from it. Similarly, you have to know exactly what the statement NOT A is so that you can work backward from it. This chapter provides rules for writing the NOT of various statements, especially those that contain quantifiers.

WRITING THE NOT OF A STATEMENT WITH QUANTIFIERS

In some instances the NOT of a statement is easy to find. For example, if A is the statement "the real number x is > 0," then the NOT of A is "it is not the case that the real number x is > 0," or equivalently, "the real number x is not > 0." The word "not" can be eliminated altogether by incorporating it into the statement to obtain "the real number x is ≤ 0."

A more challenging situation arises when the statement contains quantifiers. For instance, suppose that the statement B contains the quantifier "for all" in the standard form:

B: For all "objects" with a "certain property," "something happens."

Then the NOT of this statement is:

NOT B: It is not the case that, for all "objects" with the "certain property," "something happens"

which really means that

NOT B: There is an object with the certain property for which the something does *not* happen.

Similarly, if the statement B were to contain the quantifier "there is" in the standard form:

B: There is an "object" with the "certain property" such that "something happens,"

then the NOT of this statement is:

NOT B: It is not the case that there is an "object" with the "certain property" such that "something happens,"

or, in other words,

NOT B: For all objects with the certain property, the something does *not* happen.

In general, there are three easy steps to finding the NOT of a statement containing one or more quantifiers:

Step 1. Put the word NOT in front of the entire statement.

Step 2. If the word NOT appears to the left of a quantifier, then move the word NOT to the right of the quantifier and place it just before the something that happens. As you do so, change the quantifier to its opposite—"for all" becomes "there is" and "there is" becomes "for all."

Step 3. When *all* of the quantifiers appear to the left of the NOT, eliminate the NOT by incorporating it into the statement that appears immediately to its right.

These steps are demonstrated with the following examples.

1. For every real number $x \geq 2$, $(x^2 + x - 6) \geq 0$.

Step 1. NOT [for every real number $x \geq 2$, $(x^2+x-6) \geq 0$.]

Step 2. There is a real number $x \geq 2$ such that NOT $[(x^2+x-6) \geq 0]$.

Step 3. There is a real number $x \geq 2$ such that $(x^2+x-6) < 0$.

Note in Step 2 that, when the NOT is passed from left to right, the quantifier changes but *the certain property* (namely, $x \geq 2$) *does not.* Also, since the quantifier "for every" is changed to "there exists," it becomes necessary to replace the comma by the words "such that." If the quantifier "there exists" is changed to "for all," then the words "such that" are removed and a comma is inserted, as is illustrated in the next example.

2. There is a real number $x \geq 2$ such that $(x^2+x-6) \geq 0$.

Step 1. NOT [there is a real number $x \geq 2$ such that $(x^2+x-6) \geq 0$.]

Step 2. For all real numbers $x \geq 2$, NOT $[(x^2+x-6) \geq 0$.]

Step 3. For all real numbers $x \geq 2$, $(x^2+x-6) < 0$.

If the statement you are taking the NOT of contains nested quantifiers (see Chapter 8), then Step 2 must be performed on each quantifier, in turn, as it appears from left to right. Step 2 is repeated until *all* of the quantifiers appear to the *left* of the NOT, as demonstrated in the next two examples.

3. For every real number x between -1 and 1, there is a real number y between -1 and 1 such that $(x^2+y^2) \leq 1$.

Step 1. NOT [for every real number x between -1 and 1, there is a real number y between -1 and 1 such that $(x^2+y^2) \leq 1$.]

Step 2. There is a real number x between -1 and 1 such that, NOT [there is a real number y between -1 and 1 such that $(x^2 + y^2) \leq 1$.]

Step 3. There is a real number x between -1 and 1 such that, for all real numbers y between -1 and 1, NOT $[(x^2 + y^2) \leq 1]$.

Step 4. There is a real number x between -1 and 1 such that, for all real numbers y between -1 and 1, $(x^2 + y^2) > 1$.

4. There is a real number x between -1 and 1 such that, for all real numbers y between -1 and 1, $(x^2 + y^2) \leq 1$.

Step 1. NOT [there is a real number x between -1 and 1 such that, for all real numbers y between -1 and 1, $(x^2 + y^2) \leq 1$.]

Step 2. For all real numbers x between -1 and 1, NOT [for all real numbers y between -1 and 1, $(x^2 + y^2) \leq 1$.]

Step 3. For all real numbers x between -1 and 1, there is a real number y between -1 and 1 such that NOT $[(x^2 + y^2) \leq 1]$.

Step 4. For all real numbers x between -1 and 1, there is a real number y between -1 and 1 such that $(x^2 + y^2) > 1$.

WRITING THE NOT OF OTHER STATEMENTS HAVING A SPECIAL FORM

Another situation where you must be careful is in taking the NOT of a statement containing the words AND or OR. Just as the quantifiers are interchanged when taking the NOT of the statement, so the words AND and OR interchange. Specifically, NOT [A AND

B] becomes [NOT A] OR [NOT B]. Similarly, NOT [A OR B] becomes [NOT A] AND [NOT B]. For example:

5. NOT [($x \geq 3$) AND ($y < 2$)] becomes [($x < 3$) OR ($y \geq 2$)].
6. NOT [($x \geq 3$) OR ($y < 2$)] becomes [($x < 3$) AND ($y \geq 2$)].

When taking the NOT of a statement that already contains the word "no" or "not," the result is that the NOT "cancels" the existing "not." For example, if A is the statement:

A: There is no integer x such that $x^2 + x - 11 = 0$,

then the NOT of this statement is:

NOT A: There *is* an integer x such that $x^2 + x - 11 = 0$.

SUMMARY

The following list summarizes the rules for taking the NOT of statements that have a special form.

1. NOT [there is an object with a certain property such that something happens] becomes: for all objects with the certain property, the something does not happen.
2. NOT [for all objects with a certain property, something happens] becomes: there is an object with the certain property such that the something does not happen.
3. NOT[A AND B] becomes [(NOT A) OR (NOT B)].
4. NOT[A OR B] becomes [(NOT A) AND (NOT B)].
5. NOT[NOT A] becomes A.

In the event that a statement contains nested quantifiers, the word NOT must be processed through each quantifier, as it appears from left to right.

EXERCISES

Note: All proofs should contain an analysis of proof as well as a condensed version.

11.1. Write the NOT of each of the definitions in Exercise 5.1 on page 63, that is, take the NOT of the statement that constitutes the definition of the term being defined. For instance, the NOT of the definition in Exercise 5.1(a) would be "the real number x^* is not a maximizer of the function f if there is a real number x such that $f(x) > f(x^*)$."

11.2. Write the NOT of each of the definitions in Exercise 8.1 on page 99.

11.3. Reword the following statements so that the word "not" appears explicitly. For example, the statement "$x > 0$" can be reworded to read "x is not ≤ 0."

a. For each element x in the set S, x is in T.
b. There is an angle t between 0 and $\frac{\pi}{2}$ such that $\sin(t) = \cos(t)$.
c. For every object with a certain property, something happens.
d. There is an object with a certain property such that something happens.

11.4. If the contradiction method is used to prove each of the following statements, then what would you assume?

a. For every integer $n \geq 4$, $n! > n^2$.
b. A implies (B OR C).

c. A implies (B AND C).

d. If f is a convex function of one variable, x^* is a real number, and there is a real number $\delta > 0$ such that, for all real numbers x that satisfy the property that $|x - x^*| < \delta$, $f(x) \geq f(x^*)$, then for all real numbers y, $f(y) \geq f(x^*)$.

11.5. If the contrapositive method is used to prove each of the following propositions, then what statement(s) will you work forward from and what statement(s) will you work backward from?

a. (A AND C) implies B.

b. (A OR C) implies B.

c. If n is an even integer and m is an odd integer, then either mn is divisible by 4 or n is not divisible by 4.

11.6. Prove, by the contrapositive method, that if x is a real number that satisfies the property that, for every real number $\varepsilon > 0$, $x \geq -\varepsilon$, then $x \geq 0$.

11.7. Prove, by contradiction, that if x and y are real numbers such that $x \geq 0$, $y \geq 0$, and $x + y = 0$, then $x = 0$ and $y = 0$.

TWELVE

SPECIAL PROOF TECHNIQUES

You now have three major proof techniques to help you in proving that "A implies B": the forward–backward, the contrapositive, and the contradiction methods. In addition, when B has quantifiers, you have the choose and construction methods. There are several other special forms of statements that have well-established and usually successful proof techniques associated with them. Three of these will be developed in this chapter.

UNIQUENESS METHODS

A **uniqueness method** is used when you want to show not only that there is an object with a certain property such that something happens, but also that the object is *unique* (i.e., it is the only such object). You will know to use the uniqueness method when the statement B contains the key word "unique" (or equivalent words, such as "one and only one," for example) as well as the quantifier "there is."

In such a case, your first job is to show that the desired object does exist. This can be done either by the construction or the contradiction method. The next step is to show uniqueness in one of two standard ways. The first approach, referred to as the

direct uniqueness method, has you assume that there are two objects having the certain property and for which the something happens. If there really is only *one* such object then, using the two objects with their certain properties, the something that happens, and perhaps the information in A, you must conclude that the two objects are one and the same (i.e., that they are really equal). The forward–backward method is usually the best way to prove that they are equal. This process is illustrated in the next example.

EXAMPLE 13.

Proposition. If a, b, c, d, e, and f are real numbers such that $(ad - bc) \neq 0$, then there are unique real numbers x and y such that $(ax + by) = e$ and $(cx + dy) = f$.

Analysis of proof. The existence of the real numbers x and y was established in Example 4 on page 48 via the construction method. Here, the uniqueness will be established by the direct uniqueness method. Accordingly, you assume that (x_1, y_1) and (x_2, y_2) are two objects with the certain property and for which the something happens. In this case, that means that

A1: $ax_1 + by_1 = e$ and $cx_1 + dy_1 = f$; and

A2: $ax_2 + by_2 = e$ and $cx_2 + dy_2 = f$.

Using these four equations and the assumption that A is true, it will be shown, by the forward–backward method, that the two objects are the same, that is,

B1: $(x_1, y_1) = (x_2, y_2)$.

The key question associated with $B1$ is "How can I show that two pairs of real numbers are equal?" Using Definition 4 on page 30 for equality of ordered pairs, one answer is to show that

B2: $x_1 = x_2$ and $y_1 = y_2$

or equivalently, that

B3: $(x_1 - x_2) = 0$ and $(y_1 - y_2) = 0$.

Both of these statements are obtained from the forward process by applying some algebraic manipulations to the four equations in

$A1$ and $A2$, and by using the fact that $(ad - bc) \neq 0$, as indicated in the condensed proof that follows.

Proof of Example 13. The existence of the real numbers x and y was established in Example 4 on page 48 via the construction method. Hence only the issue of uniqueness will be addressed. To that end, assume that (x_1, y_1) and (x_2, y_2) are real numbers satisfying

1. $ax_1 + by_1 = e$,
2. $cx_1 + dy_1 = f$,
3. $ax_2 + by_2 = e$,
4. $cx_2 + dy_2 = f$.

Subtracting (3) from (1) and (4) from (2) yields

5. $[a(x_1 - x_2) + b(y_1 - y_2)] = 0$, and
6. $[c(x_1 - x_2) + d(y_1 - y_2)] = 0$.

On multiplying (5) by d and (6) by b, and then subtracting (6) from (5), it follows that

7. $[(ad - bc)(x_1 - x_2)] = 0$.

Since, by hypothesis, $(ad - bc) \neq 0$, one has $(x_1 - x_2) = 0$, and hence $x_1 = x_2$. A similar sequence of algebraic manipulations will establish that $y_1 = y_2$, and thus the uniqueness is proved. ∎

The second method for showing uniqueness, called the **indirect uniqueness method**, has you assume that there are two *different* objects having the certain property and for which the something happens. Now supposedly this cannot happen, so, by using the certain property, the something that happens, the information in A, and especially the fact that the objects are different, you must then reach a contradiction. This process is demonstrated in the next example.

EXAMPLE 14.

Proposition. If r is a positive real number, then there is a unique real number x such that $x^3 = r$.

Analysis of proof. The appearance of the quantifier "there is" in the conclusion suggests using the construction method to produce a real number x such that $x^3 = r$. This part of the proof will be omitted so that the issue of uniqueness can be addressed. To that end, suppose that

A1: x and y are two *different* real numbers (i.e., $x \neq y$) such that $x^3 = r$ and $y^3 = r$.

Using this information together with the hypothesis that r is positive, and especially the fact that $x \neq y$, a contradiction will be reached by showing that $r = 0$, which contradicts the hypothesis that r is positive.

To show that $r = 0$, one can work forward. In particular, from $A1$, since $x^3 = r$ and $y^3 = r$, it follows that

A2: $x^3 = y^3$, or, $(x^3 - y^3) = 0$.

On factoring one obtains

A3: $(x - y)(x^2 + xy + y^2) = 0$.

Here is where you can use the fact that $x \neq y$ to divide both sides of $A3$ by $(x - y)$, obtaining

A4: $(x^2 + xy + y^2) = 0$.

Thinking of $A4$ as a quadratic equation of the form $(ax^2+bx+c) = 0$ in which $a = 1$, $b = y$, and $c = y^2$, the quadratic formula states that

A5: $$x = \frac{-y \pm \sqrt{(y^2 - 4y^2)}}{2} = \frac{-y \pm \sqrt{-3y^2}}{2}.$$

Since x is real and the above formula for x requires taking the square root of $-3y^2$, it must be that

A6: $y = 0$

and if $y = 0$, then, a contradiction is reached from $A1$ because

A7: $r = y^3 = 0$.

Proof of Example 14. Only the issue of uniqueness will be addressed. To that end, assume that x and y are two different real numbers for which $x^3 = r$ and $y^3 = r$. Hence it follows that $0 = (x^3 - y^3) = (x - y)(x^2 + xy + y^2)$. Since $x \neq y$ it must be that $(x^2 + xy + y^2) = 0$. By the quadratic formula,

$$x = \frac{-y \pm \sqrt{(y^2 - 4y^2)}}{2} = \frac{-y \pm \sqrt{-3y^2}}{2}.$$

Since x is real, it must be that $y = 0$. But then $r = y^3 = 0$, thus contradicting the hypothesis that r is positive. ∎

EITHER/OR METHODS

Other special proof techniques, called **either/or methods**, arise when you come across the key words "either/or" in the form "either C or D is true," (where C and D are statements). Two different proof techniques are available depending, respectively, on whether these key words arise in the forward or backward process.

To illustrate one of these methods, called **proof by elimination**, suppose that the key words "either/or" arise in the *conclusion* of a proposition whose form is "A implies C OR D." Applying the forward–backward method, you would begin by assuming A is true and you would like to conclude that either C is true or else D is true. Suppose that you were to make the additional assumption that C is *not* true. Clearly it had better turn out that, in this case, D *is* true. Thus, with a proof by elimination, you can assume that A is true and C is false; you must then conclude that D is true, as is illustrated in the next example.

EXAMPLE 15.

Proposition. If $x^2 - 5x + 6 \geq 0$, then $x \leq 2$ or $x \geq 3$.

Analysis of proof. Recognizing the key words "either/or" in the conclusion of this proposition, you should proceed with a proof by elimination. Accordingly you can assume that

A: $(x^2 - 5x + 6) \geq 0$, and

A1: $x > 2$ (NOT C).

It is your job to conclude that

B1: $x \geq 3$ (D).

Working forward from A by factoring, it follows that

A2: $(x - 2)(x - 3) \geq 0$.

Since $x > 2$ from $A1$, you can divide both sides of $A2$ by the positive number $(x - 2)$ thus obtaining

A3: $x - 3 \geq 0$.

Adding 3 to both sides of $A3$ yields precisely $B1$ thus completing the proof.

Proof of Example 15. Assume that $(x^2 - 5x + 6) \geq 0$ and $x > 2$. It follows that $(x - 2)(x - 3) \geq 0$. Since $x > 2$, $(x - 2) > 0$, and so it must be that $x \geq 3$, as desired. ∎

Note that a proof by elimination of "A implies C OR D" can be done equally well by assuming that A is true and D is false, and then concluding that C is true. Try this approach on Example 15.

The second either/or method, called **proof by cases,** is used when the *hypothesis* contains the key words either/or in the form: "C OR D implies B." According to the forward–backward method, you can assume that C OR D is true; you must conclude that B is true. The only question is: "Is it C that is true, or is it D that is true?" Since you do not know which of these is true, you should proceed by *cases*, that is, you should do *two* proofs. In the first one you assume that C is true, and you must prove that B is true; in the second one you assume that D is true, and you must prove that B is true. This technique is illustrated in the following example. Observe how the words either/or are introduced in the middle of the proof so as to use a proof by cases.

EXAMPLE 16.

Proposition. If a is a negative real number, then $y = \frac{-b}{2a}$ is a maximizer of the function $ax^2 + bx + c$.

Analysis of Proof. Since no key words appear in the hypothesis or conclusion, the forward–backward method is used to begin the proof. Working backward, the key question is "How can I show that a point (namely, $y = \frac{-b}{2a}$) is a maximizer of a function?" Applying the definition given in Exercise 5.1(a) on page 63, it is necessary to show that

B1: for every real number x, $ay^2 + by + c \geq ax^2 + bx + c$.

Recognizing the key words "for all" in the backward process, it is appropriate to proceed with the choose method. Accordingly, you should choose

A1: a real number x

for which it must be shown that

B2: $ay^2 + by + c \geq ax^2 + bx + c$.

Subtracting $ax^2 + bx + c$ from both sides and applying simple algebraic manipulations, it must be shown that

B3: $(y - x)[a(y + x) + b] \geq 0$.

If the term $(y - x)$ were 0, then $B3$ would clearly be true. Thus you can assume that

A2: $(y - x) \neq 0$.

It is important to note here that $A2$ can be rewritten so as to contain the key words "either/or" explicitly:

A3: either $(y - x) > 0$ or $(y - x) < 0$.

At this point, recognizing the key words "either/or" in the forward process, it is time to use a proof by cases. Accordingly you should do two proofs—first assume that $(y - x) > 0$ and prove that $B3$ is true; then assume that $(y - x) < 0$ and again prove that $B3$ is true. This will now be done.

Case 1: Assume that

A4: $(y - x) > 0$.

In this case both sides of $B3$ can be divided by the positive number

$(y - x)$; thus it must be shown that

B4: $a(y + x) + b \geq 0$.

Working forward from the fact that $y = \frac{-b}{2a}$ and $a < 0$ (see the hypothesis), it follows from $A4$ that

A5: $2ax + b > 0$

and so

A6: $a(y + x) + b = ax + \frac{b}{2} = \frac{(2ax+b)}{2} > 0$.

Thus $B4$ is true, and this completes the first case.

Case 2: Now assume that

A4: $(y - x) < 0$.

In this case both sides of $B3$ can be divided by the negative number $(y - x)$, thus it must be shown that

B4: $a(y + x) + b \leq 0$.

Working forward from the fact that $y = \frac{-b}{2a}$ and $a < 0$ (see the hypothesis), it follows from $A4$ that

A5: $2ax + b < 0$

and so

A6: $a(y + x) + b = ax + \frac{b}{2} = \frac{(2ax+b)}{2} < 0$.

Thus $B4$ is true, and this completes the second case, and in fact, the entire proof.

Observe that the proof of the second case above is almost identical to that of the first case except for a reversal of sign in several places. Most mathematicians would not, in general, write the details for both cases. Rather, when they recognize the similarity of the two cases they would say: "Assume, *without loss of generality*, that Case 1 occurs...." They would then proceed to present the details for that case, omitting the second case altogether. In other words, the words "assume, without loss of generality, ..." mean that the author will present only one of the cases in detail; you will have to provide the details of the other case for yourself, as is illustrated in the following condensed proof of Example 16.

Proof of Example 16. Let x be a real number; it will be shown that $ay^2+by+c \geq ax^2+bx+c$, or equivalently, that $(y-x)[a(y+x)+b] \geq 0$. (The word "let" indicates that the choose method is being used.) This is clearly true if $y-x=0$, so assume that $(y-x) \neq 0$, but then, $(y-x) > 0$ or $(y-x) < 0$. Assume, without loss of generality, that $(y-x) > 0$. Since $a < 0$ and $y = \frac{-b}{2a}$, it follows that $[a(y+x)+b] > 0$, and so the proof is complete. ∎

Observe that with a proof by cases, two separate proofs are needed to show that "C OR D implies B"—you must prove both that "C implies B" and then that "D implies B." In contrast, with a proof by elimination, only one proof is needed to show that "A implies C OR D"—you can prove either "A AND (NOT C) implies D," or "A AND (NOT D) implies C"; either one of these two proofs by itself will suffice.

MAX/MIN METHODS

The final proof techniques to be developed in this chapter are the **max/min methods** that arises in problems dealing with maxima and minima. Suppose that S is a nonempty set of real numbers having both a largest and a smallest member. For a given real number x, you might be interested in the position of the set S relative to the number x. For instance, you might want to prove one of the following statements:

1. All of S is to the right of x (see Figure 12.1a).
2. Some of S is to the left of x (see Figure 12.1b).
3. All of S is to the left of x (see Figure 12.1c).
4. Some of S is to the right of x (see Figure 12.1d).

In mathematical problems these four statements are likely to appear, respectively, as:

a. $\min\{s : s \text{ is in } S\} \geq x$.

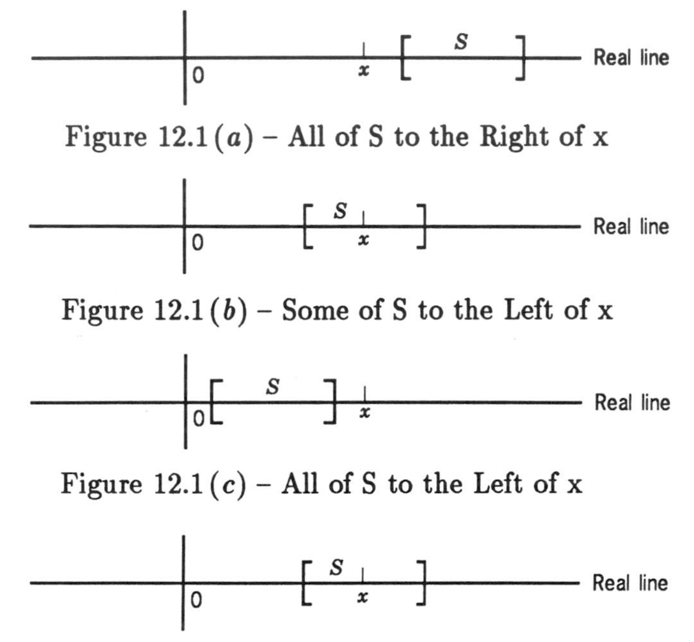

Figure 12.1 (a) – All of S to the Right of x

Figure 12.1 (b) – Some of S to the Left of x

Figure 12.1 (c) – All of S to the Left of x

Figure 12.1 (d) – Some of S to the Right of x

b. $\min\{s : s \text{ is in } S\} \leq x$.

c. $\max\{s : s \text{ is in } S\} \leq x$.

d. $\max\{s : s \text{ is in } S\} \geq x$.

The proof technique associated with the first two are discussed here and the remaining two are left as exercises. The idea behind the max/min techniques is to convert the given problem into an equivalent problem containing a quantifier; then the appropriate choose or construction method can be used.

Consider, therefore, the problem of trying to show that the smallest member of S is $\geq x$. An equivalent problem containing a quantifier can be obtained by considering the corresponding statement (1) above. Since *all* of S should be to the right of x, you need to show that, for all elements s in S, $x \leq s$, as is illustrated in the next example.

EXAMPLE 17.

Proposition. If R is the set of all real numbers, then $\min\{x(x-1) : x \text{ is in } R\} \geq -\frac{1}{4}$.

Analysis of proof. From the form of B, the max/min method will be used. As per the discussion above, the statement B can be converted to

B1: for all real numbers x, $x(x-1) \geq -\frac{1}{4}$.

Once in this form, it becomes clear that the choose method should be used to choose

A1: a real number x

for which it must be shown that

B2: $x(x-1) \geq -\frac{1}{4}$,

or equivalently, that

B3: $\left(x^2 - x + \frac{1}{4}\right) \geq 0$,

or equivalently, that

B4: $\left(x - \frac{1}{2}\right)^2 \geq 0$.

Since this is always true, the proof is complete.

Proof of Example 17. In order to conclude that $\min\{x(x-1) : x \text{ is in } R\} \geq -\frac{1}{4}$, let x be any real number. Then it follows that $x(x-1) \geq -\frac{1}{4}$ since $\left(x^2 - x + \frac{1}{4}\right) = \left(x - \frac{1}{2}\right)^2$, which is ≥ 0. ∎

Turning now to the problem of showing that the smallest member of S is $\leq x$, the approach is slightly different. To proceed, consider the corresponding statement (2) above. Since *some* of S should be to the left of x, an equivalent problem is to show that *there is* an element s in S such that $s \leq x$. Then, the construction or contradiction method could be used.

SUMMARY

This chapter has described three special proof techniques that are appropriate when statements in the forward or backward processes have the corresponding special form. Use a uniqueness method when, in the backward process, you come across the need to show that there is a *unique* object with a certain property such that something happens. With the direct uniqueness method, you should

1. Use the construction or contradiction method to establish that there is an object, say X, with the certain property and for which the something happens.
2. Assume that Y is also an object with the certain property and for which the something happens.
3. Use the properties of X and Y together with the hypothesis A to show that X and Y are the same (i.e., that $X = Y$).

With the indirect uniqueness method, you should

1. Use the construction or contradiction method to establish that there is an object, say X, with the certain property and for which the something happens.
2. Assume that Y is a *different* object from X with the certain property and for which the something happens.
3. Use the properties of X and Y, the fact that they are different, and the hypothesis A to reach a contradiction.

Use an either/or method when you encounter these key words in the forward or backward process. A proof by elimination is used to show that "A implies C OR D." To do so, follow these steps.

1. Assume that A and NOT C are true.

2. Work forward from A and NOT C to establish that D is true.
3. Work backward from D.

(You could equally well assume that A and NOT D are true, and work forward to prove that C is true. You can also work backward from C in this case.)

Use a proof by cases to show that "C OR D implies B." To do so you must do two proofs, that is,

1. Case 1: Prove that "C implies B."
2. Case 2: Prove that "D implies B."

Use a max/min method when you need to show that the largest or smallest element of a set is $\geq$ or $\leq$ some fixed number. To do so convert the statement into an equivalent statement containing a quantifier and then apply either the choose or construction method, whichever is appropriate.

EXERCISES

Note: All proofs should contain an analysis of proof as well as a condensed version.

12.1. Show that if x is a real number > 2, then there is a unique real number $y < 0$ such that $x = 2y/(1 + y)$.

12.2. If a and b are integers with $a \neq 0$ such that $a|b$, then there is a unique integer k such that $b = ka$. (See Definition 1 on page 30.)

12.3. Prove, by the indirect uniqueness method, that if m and b are real numbers with $m \neq 0$, then there is a unique number x such that $mx + b = 0$.

12.4. Prove, by the indirect uniqueness method, that there is a unique integer n for which $2n^2 - 3n - 2 = 0$.

12.5. Prove that if a and b are real numbers, at least one of which is not 0, and $i = \sqrt{-1}$, then there is a unique complex number, say $c+di$, such that $(a+bi)(c+di) = 1$.

12.6. Assuming you use the contrapositive method on each of the following problems, what proof technique will you use next? Explain.

a. A implies (C AND D).

b. (C AND D) implies B.

12.7. a. What would be the advantages and disadvantages of using the contradiction method instead of an either/or method to prove that "A implies (C OR D)"?

b. Describe how a proof by elimination would be applied to prove a statement of the form "If A then C OR D OR E."

c. Describe how a proof by cases would be applied to prove a statement of the form "If C OR D OR E then B."

12.8. Prove that if a and b are integers for which $a|b$ and $b|a$, then $a = \pm b$. (See Definition 1 on page 30.)

12.9. Prove that if m and n are integers then either 4 divides mn or 4 does not divide n.

12.10. Prove that if n and m are integers for which $n > 1$ and $n^2 + n < m$, then either n is prime or $m > 20$. (See Definition 2 on page 30.)

12.11. Consider the proposition "If x is a real number that satisfies $x^3 + 3x^2 - 9x - 27 \geq 0$, then $|x| \geq 3$."

a. Reword the proposition so that it is of the form "A implies C OR D."

b. Prove the proposition by assuming that A and NOT C are true.

c. Prove the proposition by assuming that A and NOT D are true.

12.12. Read the condensed proof in Exercise 9.12 on page 113. Explain where, why, and how a proof by cases is used.

12.13. Prove that if a, b, and c are integers for which either $a|b$ or $a|c$, then $a|(bc)$ (see Definition 1 on page 30).

12.14. Prove that if x is a real number with $|x| \geq 3$, then $x^2 - 9 \geq 0$.

12.15. Convert the following max/min problems into an equivalent statement containing a quantifier. (Note: S is a set of real numbers and x is a given real number.)

a. $\max\{s : s \text{ is in } S\} \leq x$.

b. $\max\{s : s \text{ is in } S\} \geq x$.

In the remainder of this problem, a, b, c, and u are given real numbers, and x is a variable.

c. $\min\{cx : ax \leq b \text{ and } x \geq 0\} \leq u$.

d. $\max\{cx : ax \leq b \text{ and } x \geq 0\} \geq u$.

e. $\min\{ax : b \leq x \leq c\} \geq u$.

f. $\max\{ax : b \leq x \leq c\} \leq u$.

12.16. Suppose that a, b, and c are given real numbers and that x and u are variables. Prove that $\min\{cx : ax \geq b, x \geq 0\}$ is at least as large as $\max\{ub : ua \leq c, u \geq 0\}$.

12.17. Prove that if S is a nonempty subset of a set T of real numbers and t^* is a real number such that for each element t in T, $t \geq t^*$, then $\min\{s : s \text{ is in } S\} \geq t^*$.

THIRTEEN

SUMMARY

The list of proof techniques is now complete. The techniques presented here are by no means the only ones, but they do constitute the basic set. Undoubtedly you will come across others as you are exposed to more mathematics; perhaps you will develop some of your own. In any event, there are many fine points and tricks that you will pick up with experience. A final summary of how and when to use each of the various techniques for proving the proposition "A implies B" is in order.

THE FORWARD–BACKWARD METHOD

With the forward-backward method, you can assume that A is true and your job is to prove that B is true. Through the forward process, you derive from A a sequence of statements, $A1$, $A2, \ldots$, that are necessarily true as a result of A being assumed true. This sequence is not random. It is guided by the backward process whereby, through asking and answering the key question, you derive from B a new statement, $B1$, with the property that if $B1$ is true, then so is B. This backward process can then be applied to $B1$, obtaining a new statement, $B2$, and so on. The objective is to link the forward sequence to the backward sequence by generating a statement in the forward sequence that is precisely the same as the last statement obtained in the backward sequence.

Then, like a column of dominoes, you can do the proof by going forward along the sequence from A all the way to B.

THE CONSTRUCTION METHOD

When obtaining the sequence of statements, watch for quantifiers to appear, for then the construction, choose, induction, and/or specialization methods may be useful in doing the proof. For instance, when the quantifier "there is" arises in the backward process in the standard form:

> There is an "object" with a "certain property" such that "something happens,"

consider using the construction method to produce the desired object. With the construction method, you work forward from the assumption that A is true to construct (produce, or devise an algorithm to produce, etc.) the object. However, the actual proof consists of showing that the object you constructed satisfies the certain property and also that the something happens.

THE CHOOSE METHOD

On the other hand, when the quantifier "for all" arises in the backward process in the standard form:

> For all "objects" with a "certain property," "something happens"

consider using the choose method. Here, your objective is to design a proof machine that is capable of taking any object with the certain property and proving that the something happens. To do so, you select (or choose) an object that does have the certain property. You must conclude that, for that object, the something happens. Once you have chosen the object, it is best to proceed by working forward from the fact that the chosen object does have the certain property (together with the information in A, if necessary) and backward from the something that happens.

THE INDUCTION METHOD

The induction method should be considered (even before the choose method) when the statement B has the form:

> For every integer n greater than or equal to some initial one, a statement $P(n)$ is true.

The first step of the induction method is to verify that the statement is true for the first possible value of n. The second step requires you to show that if $P(n)$ is true, then $P(n+1)$ is true. Remember that the success of a proof by induction rests on your ability to relate the statement $P(n+1)$ to $P(n)$ so that you can make use of the assumption that $P(n)$ is true. In other words, to perform the second step of the induction proof, you should write down the statement $P(n)$, replace n everywhere by $(n+1)$ to obtain $P(n+1)$, and then see if you can express $P(n+1)$ in terms of $P(n)$. Only then will you be able to use the assumption that $P(n)$ is true to reach the conclusion that $P(n+1)$ is also true.

THE SPECIALIZATION METHOD

When the quantifier "for all" arises in the forward process in the standard form:

> For all "objects" with a "certain property," "something happens"

you will probably want to use the specialization method. To do so, watch for one of these objects to arise, often in the backward process. By using specialization, you can then conclude, as a new statement in the forward process, that the something does happen for that particular object. That fact should then be helpful in reaching the conclusion that B is true. When using specialization, be sure to verify that the particular object does satisfy the certain property, for only then will the something happen.

If statements contain more than one quantifier, that is, they are nested, process them in the order in which they appear from left to right. As you read the first quantifier in the statement, identify its objects, certain property, and something that happens. Then apply an appropriate technique based on whether the statement is in the forward or backward process, and whether the quantifier is "for all" or "there is." This process is repeated until all quantifiers have been dealt with.

THE CONTRAPOSITIVE METHOD

When the original statement B contains the key word "no" or "not," or when the forward–backward method fails, you should consider the contrapositive or contradiction method. To use the contrapositive approach, write down the statements NOT B and NOT A using the techniques of Chapter 11. Then, by beginning with the assumption that NOT B is true, your job is to conclude that NOT A is true. This is best accomplished by applying the forward–backward method, working forward from the statement NOT B and backward from the statement NOT A. Remember to watch for quantifiers to appear in the forward or backward process for, if they do, then the corresponding construction, choose, induction, and/or specialization methods may be useful.

THE CONTRADICTION METHOD

In the event that the contrapositive method fails, there is still hope with the contradiction method. With this approach, you assume not only that A is true but also that B is false. This gives you two facts from which you must derive a contradiction to something that you know to be true. Where the contradiction arises is not always obvious, but it will be obtained by working forward from the statements A and NOT B.

THE UNIQUENESS METHODS

For problems containing the key words "unique," "either/or," and "max/min" you can use associated proof techniques. For example, when the conclusion of a proposition requires you to show that

> there is a unique "object" with a "certain property" such that "something happens,"

you can use the direct uniqueness method. With this method, you first establish the existence of the desired object, say X, using the construction (or contradiction) method. Then you assume that Y is a second object satisfying the certain property and for which the something happens. Using all this information (together with the hypothesis, if necessary), you must work forward to conclude that the two objects, X and Y, are the same, that is, that they are equal. You can also work backward from the fact that X is the same as Y.

With the indirect uniqueness method, you also begin by establishing the existence of the desired object X. To show uniqueness, however, you then assume that Y is a *different* object that also satisfies the certain property and the something that happens. You must use all this information, especially the fact that X and Y are different, to reach a contradiction.

THE EITHER/OR METHODS

When the key words "either/or" arise, there are two proof techniques available depending, respectively, on whether those key words appear in the forward or backward process. For example, you should use a proof by elimination when trying to prove a proposition of the form "If A then C OR D." To do so, you should assume that A is true and C is not true (that is, A and NOT C); you should then show that D is true. This can be accomplished best by using the forward–backward method. Alternatively, you can assume that A and NOT D are true; in this case you would have to show that C is true.

A proof by cases is used when trying to prove a proposition of the form "If C OR D then B." Two proofs are required. In the first case you assume that C is true and then prove that B is true; in the second case you assume that D is true and then prove that B is true.

THE MAX/MIN METHODS

When the conclusion of a proposition requires you to show that the smallest (largest) element of a set of real numbers is less (greater) than or equal to a particular real number, say x, then you should use a max/min method. Doing so involves rewriting the statement in an equivalent form using the quantifier "for all" or "there is," whichever is appropriate. Once in this form, you can then apply the choose or construction method to do the proof.

CONCLUSION

In trying to prove that "A implies B," let the form of these statements guide you as much as possible. For example, you should scan the statement B for certain key words, as they will often indicate how to proceed. If you come across the quantifier "there is," then consider the construction method, whereas the quantifier "for all" suggests using the choose or induction method. When the statement B contains the word "no" or "not," you will probably want to use the contrapositive or contradiction method. Other key words to look for are "uniqueness," "either/or," and "maximum" and "minimum," for then you would use the corresponding uniqueness, either/or, and max/min methods. If you are unable to choose an approach based on the form of B, then you should proceed with the forward–backward method. Table 5 provides a complete summary.

You are now ready to "speak" mathematics. Your new "vocabulary" and "grammar" are complete. You have learned the three major proof techniques for proving propositions, theorems, lemmas, and corollaries: the forward–backward, contrapositive, and contradiction methods. You have come to know the quantifiers and the corresponding construction, choose, induction, and specialization methods. For special situations, your bag of proof techniques includes the uniqueness, either/or, and max/min methods. If all these proof techniques fail, you may wish to stick to Greek—after all, it's all Greek to me.

Table 5. Summary of Proof Techniques

Proof Technique	When To Use It	What To Assume
Forward–Backward (page 11)	As a first attempt, or when B does not have a recognizable form	A
Contrapositive (page 115)	When B has the word "no" or "not" in it	NOT B
Contradiction (page 104)	When B has the word "not" in it, or when the first two methods fail	A and NOT B
Construction (page 47)	When B has the term "there is," "there exists," etc.	A
Choose (page 57)	When B has the term "for all," "for each," etc.	A, and choose an object with the certain property
Induction (page 69)	When B is true for each integer beginning with an initial one, say n_0	The statement is true for n

What To Conclude	How To Do It
B	Work forward from A and apply the backward process to B.
NOT A	Work forward from NOT B and backward from NOT A.
Some contradiction	Work forward from A and NOT B to reach a contradiction.
There is the desired object	Guess, construct, etc., the object. Then show that it has the certain property and that the something happens.
That the something happens	Work forward from A and the fact that the object has the certain property. Also work backward from the something that happens.
The statement is true for $n+1$. Also show it true for n_0.	First substitute n_0 everywhere and show it true. Then invoke the induction hypothesis for n to prove it true for $n+1$.

Table 5. Summary of Proof Techniques (continued)

Proof Technique	When To Use It	What To Assume
Specialization (page 83)	When A has the term "for all," "for each," etc.	A
Direct Uniqueness (page 134)	When B has the word "unique" in it	There are two such objects, and A
Indirect Uniqueness (page 135)	When B has the word "unique" in it	There are two different objects, and A
Proof by Elimination (page 137)	When B has the form "C OR D"	A and NOT C or A and NOT D
Proof by Cases (page 138)	When A has the form "C OR D"	Case 1: C Case 2: D
Max/min 1 (page 141)	When B has the form "max $S \leq x$" or "min $S \geq x$"	Choose an s in S, and A
Max/min 2 (page 141)	When B has the form "max $S \geq x$" or "min $S \leq x$"	A

What To Conclude	How To Do It
B	Work forward by specializing A to one particular object having the certain property.
The two objects are equal	Work forward using A and the properties of the objects. Also work backward to show the objects are equal.
Some contradiction	Work forward from A using the properties of the two objects and the fact that they are different.
D	Work forward from A and NOT C, and backward from D
or	or
C	Work forward from A and NOT D, and backward from C.
B B	First prove that C implies$B$; then prove that D implies B.
$s \leq x$ or $s \geq x$	Work forward from A and the fact that s is in S. Also work backward.
Construct s in S so that $s \geq x$ or $s \leq x$	Use A and the construction method to produce the desired s in S

EXERCISES

13.1. For each of the following statements, indicate which proof technique you would use to begin the proof and explain why.

a. If p and q are odd integers then the equation $x^2 + 2px + 2q = 0$ has no rational solution for x.

b. For every integer $n \geq 4$, $n! > n^2$.

c. If f and g are convex functions, then $f+g$ is a convex function.

d. If a, b, and c are real variables, then the maximum value of $ab + bc + ca$ subject to the condition that $a^2 + b^2 + c^2 = 1$ is 1.

e. In a plane, there is one and only one line perpendicular to a given line ℓ through a point P on the line.

f. If f and g are two functions such that (1) for all real numbers x, $f(x) \leq g(x)$ and (2) there is no real number M such that, for all x, $f(x) \leq M$, then there is no real number $M > 0$ such that, $\forall x$, $g(x) \leq M$.

g. If f and g are continuous functions at the point x, then so is the function $f + g$.

h. If f and g are continuous functions at the point x, then for every real number $\varepsilon > 0$ there is a real number $\delta > 0$ such that, for all real numbers y with $|x - y| < \delta$, $|f(x) + g(x) - (f(y) + g(y))| < \varepsilon$.

i. If f is the function defined by $f(x) = 2^x - \frac{x^2}{2}$, then there is a real number x^* between 0 and 1 such that, for all y, $f(x^*) \leq f(y)$.

13.2. For each of the problems in Exercise 13.1, state how the technique you chose to begin the proof would be applied to that problem. That is, indicate what you would assume, what you would conclude, and how you go about doing it.

13.3. Describe how you would use each of the following proof techniques to prove that "for every integer $n \geq 4$, $n! > n^2$." State what you would assume and what you would conclude.

a. Induction method.

b. Choose method.

c. Forward–Backward method. (Hint: Convert the problem to an equivalent one of the form "if ... then ....")

d. Contradiction method.

13.4. Suppose the forward–backward method is used to start each of the following proofs. Make a list of all proof techniques that are likely to be used subsequently in the proof. Repeat this exercise assuming the contradiction method is used to start the proof.

a. (C AND D) implies (E OR F).

b. (C OR D) implies (E AND F).

c. If X is an object such that for all objects Y with a certain property, something happens, then there is an object Z with a certain property such that something else happens.

d. If for all objects X with a certain property, something happens, then there is an object Y with a certain property such that, for all objects Z with a certain property, something else happens.

APPENDIX A:

PUTTING IT ALL TOGETHER: PART I

Through the use of a specific example, this appendix will show you how to read and how to understand a written proof as it might appear in a textbook. Some general problem-solving strategies will also be discussed and demonstrated on a problem.

HOW TO READ A PROOF

Due to the way in which most proofs are currently written, it is necessary for you to figure out which proof techniques are being used and why. There are three reasons why proofs are often difficult to read:

1. The steps of a proof are not necessarily presented in the order in which they were performed when doing the proof.
2. The author does not often refer to the techniques by name.
3. Several individual steps are usually combined into a single sentence with little or no justification.

In order to read a proof, you will have to reconstruct the author's thought process. Doing so requires that you identify which proof techniques are being used and how they apply to the particular problem. The next example demonstrates how to read a proof. Unlike previous examples in this book, the condensed version of the proof precedes the analysis, which then explains how to read the condensed proof.

The example deals with the concept of an "unbounded" set of real numbers, meaning that there are elements in the set that are "arbitrarily" far away from 0. A formal definition of a bounded set will be given first.

Definition 19. A set S of real numbers is **bounded** if and only if there is a real number $M > 0$ such that, for all elements x in S, $|x| < M$.

EXAMPLE 18.

Proposition. If S is a subset of a set T of real numbers and S is not bounded, then T is not bounded.

Proof of Example 18. (For reference purposes, each sentence of the proof is written on a separate line.)

S1. Suppose that S is a subset of T and assume to the contrary that T is bounded.

S2. Hence there is a real number $M' > 0$ such that, for all x in T, $|x| < M'$.

S3. It will be shown that S is bounded.

S4. To that end, let x' be an element of S.

S5. Since S is a subset of T, it follows that x' is in T.

S6. But then $|x'| < M'$ and so S is bounded, which is a contradiction. ∎

Analysis of Proof. An interpretation of each of the statements $S1$ through $S6$ is given below.

Interpretation of S1. The author is indicating that the proof is going to be done by contradiction and appropriately is assuming that (part of) A as well as NOT B are true, that is, that

A1: S is a subset of T, and

A2: T is bounded. (NOT B)

Note that the author has not explicitly stated that S is not bounded, although it is being assumed true.

The use of the contradiction method should not come as a surprise since the conclusion of the example contains the word "not." Sometimes it is beneficial to scan through the proof to find the contradiction so that you will know to what end the author is working forward from A and NOT B. In $S6$ it says that S is bounded, and this contradicts the hypothesis that S is not bounded. Statements $S2$ through $S6$ should indicate how the author has worked forward from A and NOT B to reach the contradiction.

Interpretation of S2. The author is working forward from the statement NOT B (i.e., that T is a bounded set) via Definition 19 to claim that

A3: there is the real number $M' > 0$ such that, for all x in T, $|x| < M'$.

Interpretation of S3. It is here that the author states the desired objective of showing that

B1: S is bounded.

Observe that there is no indication that the reason for wanting to show that S is bounded is to reach a contradiction.

Interpretation of S4. This statement might seem cryptic at first, and the reason is that the author has omitted several steps of the thought process. What has happened is that the author has implicitly posed the key question associated with $B1$: "How can I show that a set (namely, S) is bounded?" and has then used Definition 19 to answer it, whereby it must be shown that

B2: there is a real number $M > 0$ such that, for all x in S, $|x| < M$.

Recognizing that the first quantifier appearing in $B2$ is "there is," the construction method should be used to produce the desired value for M. What the author has failed to tell you is that the value for M is M' (see $A3$), that is, the author has constructed

A4: $M = M'$.

Given that this is the case, according to the construction method, it must be shown that M has the desired properties, that is, that

B3: $M > 0$ (which it is since $M = M'$ and $M' > 0$) and

B4: for all x in S, $|x| < M = M'$.

Recognizing the appearance of the quantifier "for all" in the backward process ($B4$), one should then proceed by the choose method. This is precisely what the author has done in $S4$, where it states: " ...let x' be an element of S," that is, the author has chosen

A5: an element x' in S.

According to the choose method, it must then be shown that

B5: $|x'| < M = M'$,

which the author has not explicitly said would be done.

What makes $S4$ so difficult to understand is the lack of sufficient detail, particularly the omission of the key question and answer, and the combining of several steps into one sentence. Such details are often left for you to decipher for yourself.

Interpretation of S5. In $S5$, the author has suddenly switched to the forward process and is working forward from A (i.e., from the fact that S is a subset of T) by using Definition 14 of a subset (see page 55) to conclude that

A6: for all elements x in S, x is in T.

Then, without stating so, the author has specialized this statement to the particular value of x' in S that was chosen in $A5$, thus concluding that

A7: x' is in T.

Note once again that part of the author's thought process has been omitted and left for you to figure out. At this point it is not clear why it is necessary to show that x' is in T. The reason should be related to showing that $|x'| < M'$ (see $B5$).

Interpretation of S6. It is here that the author finally concludes that $|x'| < M'$, thus showing that S is bounded and reaching a contradiction to the hypothesis that S is not bounded. The only question is *how* the author reached the conclusion that $|x'| < M'$. Once again the specific proof technique has not been mentioned. Actually, the author has used specialization. Specifically, the for-all statement in $A3$ has been specialized to the particular value of x'. However, when using specialization, recall that it is necessary to verify that the particular object under consideration (x', in this case) has the certain property in the for-all statement (that x' is in T). The fact that x' is in T was established in $S5$, but observe that, at the time, the author did not tell you why it was being shown that x' is in T.

Read the condensed proof again. Observe that what makes it difficult to understand is the lack of reference to the specific techniques that were used, especially the backward process and the specialization method. The author has also omitted part of the thought process. To understand the proof you have to fill in the missing steps.

In summary, when reading proofs, expect to do some work on your own. Learn to identify the various techniques that are being used. Begin by trying to determine if the forward–backward or contradiction method is the primary technique. Then try to follow the methodology associated with that technique. Be watchful for quantifiers to appear, for then the corresponding choose, induction, construction, and/or specialization methods are likely to be used. Be particularly careful of notational complications that can arise with the choose and specialization methods. For instance, in the condensed proof of Example 18, the author could have used the symbols x and M instead of x' and M'. If this had been done, then you would have had to distinguish between such statements as "for all x with a certain property, something happens," in which

"x" refers to a general object, and a statement of the form "let x have a certain property," in which "x" refers to a specific object. The double use of a symbol is confusing but common in proofs, as is illustrated in the example in Appendix B.

When you are unable to follow a particular step of a written proof, it is most likely due to the lack of sufficient detail. To fill in the gaps, learn to ask yourself how *you* would proceed to do the proof. Then try to see if the written proof matches your thought process.

HOW TO DO A PROOF

Now that you know how to read a proof, it is time to see how to do a proof on your own. In the example that follows, pay particular attention to how the form of the statement under consideration dictates the technique to be used.

EXAMPLE 19.

Proposition. If T is a bounded subset of real numbers, then any subset of T is bounded.

Analysis of proof. Looking at the statement B, you should recognize the quantifier "for any" and thus you should begin with the choose method. Accordingly, you would write: "Let S be a subset of T. It will be shown that S is bounded." In other words, you should choose

A1: a subset S of T

for which it must be shown that

B1: S is bounded.

Not recognizing any special key words in $B1$, it is probably best to proceed with the forward–backward method. The idea is to work backward from $B1$ until it is no longer fruitful to do so. Then the forward process will be applied to the hypothesis that

T is a bounded set of real numbers.

Working backward from $B1$, can you pose a key question? One such question is "How can I show that a set (namely, S) is bounded?" Using Definition 19 means you must show that

B2: there is a real number $M > 0$ such that, for all elements x in S, $|x| < M$.

The appearance of the quantifier "there is" in $B2$ suggests using the construction method and also indicates that you should turn to the forward process to produce the desired value for M. Working forward from the hypothesis that T is bounded, is there anything that you can say is true? At least one possibility is to use Definition 19, whereby you can conclude that

A2: there is a real number $M' > 0$ such that, for all elements x in T, $|x| < M'$.

Observe that the symbol M' has been used instead of M because M was already used in the backward process.

Perhaps M' is the desired value of M. When using the construction method, it is generally advisable to try to construct the desired object in the most "obvious" way. In this case, that means setting

A3: $M = M'$.

If this "guess" is correct, then you must still show that the certain property holds (i.e., $M > 0$) and also that the something happens (i.e., that for all x in S, $|x| < M$). It is not hard to see that $M > 0$ because $M' > 0$ and $M = M'$. Thus it remains only to show that

B3: for all elements x in S, $|x| < M = M'$.

The appearance of the quantifier "for all" in $B3$ suggests proceeding with the choose method, whereby you would choose

A4: an element x' in S

for which you must show that

B4: $|x'| < M$, or, since $M = M'$, $|x'| < M'$.

Observe that the symbol x' has been used for the particular object to avoid confusion with the general symbol x. At this point it might not be clear how to reach the conclusion that $|x'| < M'$, but look carefully at the for-all statement in $A2$. You can use specialization to reach the desired conclusion that $|x'| < M'$, provided that you can specialize the statement to $x = x'$. In order to be able to do so, you must make sure that the particular object under consideration (namely, x') does satisfy the certain property in the for-all statement (namely, that x' is in T). Your objective, therefore, is to show that

B5: x' is in T.

Recall that x' was chosen to be an element of S (see $A4$) and that S was chosen to be a subset of T (see $A1$). How can you use this information to show that x' is in T? Working forward from Definition 14 on page 55, you know that

A5: for all x in S, x is in T.

Thus, you can specialize this statement to $x = x'$, which you know is in S (see $A4$), and reach the desired conclusion that x' is in T, completing the proof.

Proof of Example 19. Let S be a subset of T. To show that S is bounded, a real number $M > 0$ will be produced with the property that, for all x in S, $|x| < M$.

By hypothesis, T is bounded, and so there is a real number $M' > 0$ such that, for all x in T, $|x| < M'$. Set $M = M'$. To see that M has the desired properties, let x' be an element of S. Since S is a subset of T, x' is in T. But then it follows that $|x'| < M' = M$, as desired. ∎

Problem solving is not a precise science. Some general suggestions will be given here, but a more detailed discussion can be found in G. Polya's book *How to Solve It* (Princeton University Press, Princeton, NJ, 1945) or in W. Wickelgren's book *How to Solve Problems* (W. H. Freeman, San Francisco, CA, 1974).

When trying to prove that "A implies B," consciously choose a technique based on the form of B. If no key words are apparent, then it is probably best to proceed with the forward–backward method. If unsuccessful, there are several avenues to pursue before giving up. You might try asking yourself why B cannot be false, thus leading you into the contradiction (or contrapositive) method.

The point is that, when you are unable to complete a proof for one reason or another, it is advisable to seek another proof technique consciously. Sometimes the form of the statement B can be manipulated so as to induce a different technique. For instance, suppose that the statement B contains the quantifier "for all" in the standard form:

> For all "objects" with a "certain property," "something happens."

If the choose method fails, then you might try rewriting B so that it reads:

> There is no "object" with the "certain property" such that the "something" does not happen.

Now the contradiction method suggests itself because the key word "no" appears. Thus you would assume that there is an object with the certain property such that the something does not happen. Your objective would be to reach a contradiction.

When the statement B has recognizable key words, it is best to try the corresponding proof technique first. Remember that, as you proceed through a proof, different techniques will be needed as the form of the statement under consideration changes.

If you are really stuck, it can sometimes be advantageous to leave the problem for a while, for when you return to it, you might see a new approach. Undoubtedly you will learn many tricks of your own as you solve more and more problems.

APPENDIX B:

PUTTING IT ALL TOGETHER: PART II

This appendix will give you more practice in reading and doing proofs. The examples presented here are more difficult than those in Appendix A. The additional difficulty is due, in part, to the fact that the statements under consideration contain three nested quantifiers as opposed to two.

HOW TO READ A PROOF

The first example is designed to teach you how to read a condensed proof as it might appear in a textbook. As such, the condensed version of the proof precedes the detailed analysis which then explains how to read the condensed proof. The example deals with the concept of a "continuous" function of one variable, meaning that if the value of the variable is changed "slightly," then the value of the function does not change "radically." Another way of saying that a function is continuous is that its graph can be drawn without lifting the pencil. To proceed formally, a mathematical definition of continuity is needed.

Definition 20. A function f of one variable is **continuous at the**

> **point x** if and only if for each real number $\varepsilon > 0$, there is a real number $\delta > 0$ such that, for all real numbers y with the property that $|x - y| < \delta$, it follows that $|f(x) - f(y)| < \varepsilon$.

EXAMPLE 20.

Proposition. If f and g are two functions that are continuous at x, then the function $f + g$ is also continuous at x. [Note: $f + g$ is the function whose value at any point y is $f(y) + g(y)$].

Proof of Example 20. (For reference purposes, each sentence of the proof is written on a separate line.)

S1. To see that the function $f+g$ is continuous at x, let $\varepsilon > 0$.

S2. It will be shown that there is a $\delta > 0$ such that, for all y with $|x - y| < \delta$, $|f(x) + g(x) - [f(y) + g(y)]| < \varepsilon$.

S3. Since f is continuous at x, there is a $\delta_1 > 0$ such that, for all y with $|x - y| < \delta_1$, $|f(x) - f(y)| < \frac{\varepsilon}{2}$.

S4. Similarly, since g is continuous at x, there is a $\delta_2 > 0$ such that, for all y with $|x - y| < \delta_2$, $|g(x) - g(y)| < \frac{\varepsilon}{2}$.

S5. Let $\delta = \min\{\delta_1, \delta_2\}$ (which is > 0) and let y have the property that $|x - y| < \delta$.

S6. Then $|x - y| < \delta_1$ and $|x - y| < \delta_2$, and so

$$\begin{aligned} |f(x) + g(x) - [f(y) + g(y)]| &\leq |f(x) - f(y)| + \\ &\quad\ |g(x) - g(y)| \\ &< \tfrac{\varepsilon}{2} + \tfrac{\varepsilon}{2} \\ &= \varepsilon. \ \blacksquare \end{aligned}$$

Analysis of proof. An interpretation of each of the statements $S1$ through $S6$ is given below.

Interpretation of S1. This statement indicates that the forward–backward method is going to be used because the statement "to see that $f + g$ is continuous at x" means that the author is about to show that the statement B is true. The question is why did the author say "...let $\varepsilon > 0$?" Hopefully it is related to showing that $f + g$ is continuous at x. What has happened is that the author has implicitly posed the key question "How can I show that a function (namely, $f + g$) is continuous at a point (namely, x)?" and has used Definition 20 to answer it, so it must be shown that

B1: for every $\varepsilon > 0$, "something happens."

Recognizing the appearance of the quantifier "for every" in the backward process, one should then proceed with the choose method, whereby you would choose

A1: a real number $\varepsilon > 0$.

The author has done this when it says "...let $\varepsilon > 0$." Observe that the symbol ε appearing in the definition of continuity refers to a *general* value for ε while the symbol ε appearing in $S1$ refers to a *specific* value resulting from the choose method. The double use of a symbol can be confusing but is common in written proofs.

What makes $S1$ so difficult to understand is that the author has omitted part of the thought process, namely, the key question and answer. Such details are often left for you to figure out.

Interpretation of S2. The author is stating precisely what is going to be done, but why should the author want to show that there is a $\delta > 0$ such that ...? Recall that, in $S1$, the choose method was used to select an $\varepsilon > 0$ (see $A1$) and so it must be shown that, for that ε, the something happens. That something is

B2: there is a $\delta > 0$ such that for all y with $|x - y| < \delta$,
$|f(x) + g(x) - [f(y) + g(y)]| < \varepsilon$.

as stated in $S2$ (see Definition 20).

Interpretation of S3. The author has turned to the forward process (having recognized from $B2$ that the quantifier "there is" suggests using the construction method to produce the desired δ). The author is working forward from the hypothesis that f is continuous at x by using Definition 20. What the author has failed to tell you is that, in the definition, the quantifier "for all" arises, and the statement has been specialized to the value of $\frac{\varepsilon}{2}$, where ε is the one that was chosen in $A1$, thus leading to

A2: there is a $\delta_1 > 0$ such that, for all y with $|x - y| < \delta_1$, $|f(x) - f(y)| < \frac{\varepsilon}{2}$.

At this moment, it is not clear why the author specialized to the value of $\frac{\varepsilon}{2}$ as opposed to ε. The reason will become clear in $S6$.

Interpretation of S4. The author is working forward from the fact that g is continuous at x in precisely the same way that was done in $S3$. Thus the author can conclude that

A3: there is a $\delta_2 > 0$ such that, for all y with $|x - y| < \delta_2$, $|g(x) - g(y)| < \frac{\varepsilon}{2}$.

Keep in mind that, from $B2$, the author is trying to construct a value of $\delta > 0$ for which something happens.

Interpretation of S5. It is here that the desired value of $\delta > 0$ is produced. Specifically, the values of δ_1 and δ_2 from $A2$ and $A3$, respectively, are combined to give the desired δ, namely,

A4: $\delta = \min\{\delta_1, \delta_2\}$.

It is not yet clear why δ has been constructed in this way. In any event, it is the author's responsibility to show that $\delta > 0$ and also that δ satisfies the something that happens associated with the quantifier "there is" in $B2$, i.e.,

B3: for all y with $|x - y| < \delta$, $|f(x) + g(x) - [f(y) + g(y)]| < \varepsilon$.

In $S5$ the author remarks that $\delta > 0$ (which is true because both δ_1 and δ_2 are > 0). Also, the reason that the author has said "...let y have the property that $|x - y| < \delta$" is because the choose method is being used to show that $B3$ is true. Thus, the author has chosen

A5: a real number y with $|x - y| < \delta$.

Note again the double use of the symbol y. Nonetheless, to complete the proof, it must be shown that

B4: $|f(x) + g(x) - [f(y) + g(y)]| < \varepsilon$.

Interpretation of S6. It is here that the author concludes that $B4$

is true. The only question is how? It is not hard to see that

$$\textbf{A6:} \quad |f(x)+g(x)-[f(y)+g(y)]| \leq |f(x)-f(y)|+ |g(x)-g(y)|$$

because the absolute value of the sum of two numbers—namely, $f(x)-f(y)$ and $g(x)-g(y)$—is always $\leq$ the sum of the absolute values of the two numbers. But why is $|f(x)-f(y)|+|g(x)-g(y)| < \frac{\varepsilon}{2}+\frac{\varepsilon}{2}$? Once again the author has omitted part of the thought process.

What has happened is that specialization has been used to claim that $|f(x)-f(y)| < \frac{\varepsilon}{2}$ and $|g(x)-g(y)| < \frac{\varepsilon}{2}$. Specifically, the for-all statements in $A2$ and $A3$ have both been specialized to the particular value of y that was chosen in $A5$. Recall that, when using specialization, one must be sure that the particular object under consideration (namely, y) satisfies the certain property (in this case, $|x-y| < \delta_1$ and $|x-y| < \delta_2$). Note that, in $S6$, the author does claim that $|x-y| < \delta_1$ and $|x-y| < \delta_2$, but can you see *why* this is true? Look at $A4$ and $A5$. Since y was chosen so that $|x-y| < \delta$ and $\delta = \min\{\delta_1, \delta_2\}$, it must be that both $\delta \leq \delta_1$ and $\delta \leq \delta_2$, so

$$\textbf{A7:} \quad |x-y| < \delta_1 \quad \text{and} \quad |x-y| < \delta_2.$$

Hence, the author is justified in specializing the for-all statements in $A2$ and $A3$ to this particular value of y and hence can conclude that

$$\textbf{A8:} \quad |f(x)-f(y)| < \tfrac{\varepsilon}{2} \text{ and}$$

$$\textbf{A9:} \quad |g(x)-g(y)| < \tfrac{\varepsilon}{2}.$$

In other words, the author is justified in claiming that

$$\begin{aligned} |f(x)+g(x)-[f(y)+g(y)]| &\leq |f(x)-f(y)|+|g(x)-g(y)| \\ &< \tfrac{\varepsilon}{2}+\tfrac{\varepsilon}{2} \\ &= \varepsilon. \end{aligned}$$

Perhaps now it is clear why the author specialized the definitions of continuity for f and g to $\frac{\varepsilon}{2}$ instead of ε (see $A2$ and $A3$).

Again, note that what makes $S6$ so hard to understand is the omission of part of the thought process. You must learn to fill in the missing links by determining which proof techniques are being used and consequently what has to be done.

HOW TO DO A PROOF

The next example illustrates how to go about doing your own proof. Pay particular attention to how the form of the statement under consideration leads to the correct technique.

EXAMPLE 21.

Proposition. If f is a function that, at the point x, satisfies the property that there are real numbers $c > 0$ and $\delta' > 0$ such that, for all y with $|x - y| < \delta'$, $|f(x) - f(y)| \leq c|x - y|$, then f is continuous at x.

Analysis of proof. Begin by looking at the statement B and trying to select a proof technique. Since B does not have a particular form, the forward–backward method should be used. Thus you are led to the key question "How can I show that a function (namely, f) is continuous at a point (namely, x)?" As usual, a definition is available to provide an answer, that being to show that

B1: for all $\varepsilon > 0$, there is a $\delta > 0$ such that, for all y with $|x - y| < \delta$, $|f(x) - f(y)| < \varepsilon$.

Looking at $B1$, you should recognize the first "for all" as the first of the nested quantifiers. Hence you should use the choose method to choose

A1: a real number $\varepsilon' > 0$.

It is advisable to use a symbol other than ε for the specific choice so as to avoid confusion. In the condensed proof you would write "Let $\varepsilon' > 0$...." Once you have selected $\varepsilon' > 0$, the choose method requires you to show that the something happens, in this case that

B2: there is a $\delta > 0$ such that, for all y with $|x - y| < \delta$, $|f(x) - f(y)| < \varepsilon'$.

Looking at $B2$, you should recognize the key words "there is" as the next nested quantifier. Accordingly you should use the construction method to produce the desired value of δ. When the construction method is used, it is advisable to turn to the forward process to produce the desired object.

Working forward from the hypothesis that there are real numbers $c > 0$ and $\delta' > 0$ for which something happens, you should attempt to construct δ. Perhaps $\delta = \delta'$ (or perhaps $\delta = c$). Why not "guess" that

A2: $\delta = \delta'$.

If this guess is not correct, then maybe you will discover what the proper value for δ should be. Alternatively, you could attempt to construct δ by working backward from the fact that you want δ to satisfy the property that, for all y with $|x - y| < \delta$, $|f(x) - f(y)| < \varepsilon'$.

In any event, to check if the current guess of $\delta = \delta'$ is correct, it is necessary to see if the certain property associated with the quantifier "there is" in $B2$ holds for δ' and also if the something happens. From $B2$, the certain property is that $\delta > 0$, but since $\delta = \delta'$ and $\delta' > 0$, it must be that $\delta > 0$. It remains only to verify that, for $\delta = \delta'$,

B3: for all y with $|x - y| < \delta$, $|f(x) - f(y)| < \varepsilon'$.

Looking at $B3$, you should recognize the quantifier "for all" in the backward process and should use the choose method to select

A3: a real number y' with $|x - y'| < \delta$

for which it must be shown that

B4: $|f(x) - f(y')| < \varepsilon'$.

To see if $B4$ is true, the hypothesis still has some unused information. Specifically, the number $c > 0$ has not been used, nor has the statement:

A4: for all y with $|x - y| < \delta'$, $|f(x) - f(y)| \leq c|x - y|$.

Upon recognizing the quantifier "for all" in the forward process (in $A4$), you should consider using specialization; the only question is which value of y to specialize to. Why not try the value of y' chosen in $A3$? Fortunately y' was chosen to have the certain property in $A4$ (i.e., $|x - y'| < \delta$ and $\delta = \delta'$, so $|x - y'| < \delta'$), and so you can specialize $A4$ to y' obtaining:

A5: $|f(x) - f(y')| \leq c|x - y'|$.

Recall that the last statement obtained in the backward process was $B4$, so the idea is to see if $A5$ can be made to look more like $B4$. For instance, since y' was chosen in $A3$ so that $|x - y'| < \delta$, $A5$ can be rewritten as

A6: $|f(x) - f(y')| \leq c\delta$.

$B4$ would be obtained if $c\delta$ were known to be $< \varepsilon'$. While from $A2$ you know that $\delta = \delta'$, you do *not* know that $c\delta' < \varepsilon'$. This means that the original guess of $\delta = \delta'$ in $A2$ was incorrect. Perhaps δ should have been chosen > 0 and with the property that $c\delta < \varepsilon'$, or equivalently, since $c > 0$, why not construct

A2: $0 < \delta < \frac{\varepsilon'}{c}$.

To see if the new "guess" of $0 < \delta < \frac{\varepsilon'}{c}$ is correct, it is necessary to check if, for this value of δ, the certain property associated with the quantifier "there is" in $B2$ holds and also whether the something happens. Clearly δ has been chosen to be > 0 and thus has the certain property. It remains only to verify that the something happens, in this case, that $B3$ is true.

As before, one would proceed with the choose method to select y' with $|x - y'| < \delta$ (see $A3$), and again one is led to showing that $B4$ is true. The previous approach was to specialize $A4$ to y'. This time, however, there is a problem because y' is *not* known to satisfy the certain property in $A4$: that $|x - y'| < \delta'$. Unfortunately, all you do know is that $|x - y'| < \delta$ (see $A3$) and $0 < \delta < \frac{\varepsilon'}{c}$ (see the "new" $A2$). If only you could apply specialization to $A4$, then you would obtain $A5$, and hence $A6$. Finally, since $c\delta < \varepsilon'$, you could reach the desired conclusion that $B4$ is true.

Can you figure out how to choose δ so that both $c\delta < \varepsilon'$ and so that $A1$ can be specialized to y'? The answer is to construct

A2: $0 < \delta < \min\{\delta', \frac{\varepsilon'}{c}\}$, for then, when y' is chosen with $|x-y'| < \delta$ in $A3$, it will follow that $|x - y'| < \delta'$ because $\delta \leq \delta'$. Thus, it will be possible to specialize $A4$ to y'. Also, from $A5$ and the fact that $\delta \leq \frac{\varepsilon'}{c}$, it will be possible to conclude that $|f(x) - f(y')| \leq c|x - y'| \leq c\delta < c(\frac{\varepsilon'}{c}) = \varepsilon'$, thus completing the proof.

In the condensed proof that follows, observe that the construction of the correct value for δ is presented at the beginning of the proof—in the second sentence—which does not correspond to the order in which this value was constructed in the analysis given above.

Proof of Example 21. To show that f is continuous at x, let $\varepsilon' > 0$. By the hypothesis that $c > 0$, it is possible to construct $0 < \delta < \min\{\delta', \frac{\varepsilon'}{c}\}$. Then, for y' with $|x - y'| < \delta$, it follows that $|x - y'| < \delta'$ and so, from the hypothesis, $|f(x) - f(y')| \leq c|x - y'| \leq c\delta$. Furthermore, since $\delta < \frac{\varepsilon'}{c}$, $|f(x) - f(y')| \leq c\delta < c(\frac{\varepsilon'}{c}) = \varepsilon'$, and the proof is complete. ■

As with any language, reading, writing, and speaking come only with practice. The proof techniques are designed to get you started in the right direction.

SOLUTIONS TO EXERCISES

SOLUTIONS TO CHAPTER 1

1.1 (a), (c), (e), and (f) are statements.

1.3
a. Hypothesis: The right triangle XYZ with sides of lengths x and y, and hypotenuse of length z, has an area of $\frac{z^2}{4}$.
Conclusion: The triangle XYZ is isosceles.

b. Hypothesis: n is an even integer.
Conclusion: n^2 is an even integer.

c. Hypothesis: a, b, c, d, e, and f are real numbers that satisfy the property that $ad - bc \neq 0$.
Conclusion: The two linear equations $ax + by = e$ and $cx + dy = f$ can be solved for x and y.

d. Hypothesis: n is a positive integer.
Conclusion: The sum of the first n integers is $n(n+1)/2$.

e. Hypothesis: r is a real number and satisfies $r^2 = 2$.
Conclusion: r is an irrational number.

f. Hypothesis: p and q are positive real numbers with $\sqrt{pq} \neq (p+q)/2$.
Conclusion: $p \neq q$.

g. Hypothesis: x is a real number.

Conclusion: The minimum value of $x(x-1)$ is at least $-\frac{1}{4}$.

1.5 a. True because A is false.

b. True because A and B are both true.

c. True because B is true (the truth of A does not matter).

d. True if $x \neq 3$ because then A is false.
False if $x = 3$ because then A is true and B is false.

1.7 If you want to prove that "A implies B" is true and you know that B is false, then A should also be false. The reason is that if A is false, then it does not matter whether B is true or false, and Table 1 ensures that "A implies B" is true. On the other hand, if A is true and B is false, then "A implies B" would be false.

1.9 a. (T = true, F = false)

A	B	C	$(B \Rightarrow C)$	$A \Rightarrow (B \Rightarrow C)$
T	T	T	T	T
T	T	F	F	F
T	F	T	T	T
T	F	F	T	T
F	T	T	T	T
F	T	F	F	T
F	F	T	T	T
F	F	F	T	T

b. (T = true, F = false)

A	B	C	$(A \Rightarrow B)$	$(A \Rightarrow B) \Rightarrow C$
T	T	T	T	T
T	T	F	T	F
T	F	T	F	T
T	F	F	F	T
F	T	T	T	T
F	T	F	T	F
F	F	T	T	T
F	F	F	T	F

SOLUTIONS TO CHAPTER 2

2.1 The forward process is a process that makes specific use of the information contained in statement A, the hypothesis. The backward process is a process that tries to find a chain of statements leading to the fact that statement B, the conclusion, is true.

The backward process starts with the statement B that you are trying to conclude is true. By asking and answering key questions, you derive a sequence of new statements with the property that if the sequence of new statements is true, then B is true. The backward process continues in this manner either until you obtain the statement A, or until you can no longer ask and/or answer the key question.

The forward process begins with the statement A, which

is assumed to be true, and derives from it a sequence of new statements that are true as a result of statement A being true. Every new statement derived from A should be directed toward linking up with the last statement obtained in the backward process. The last statement of the backward process acts as the guiding light in the forward process, just as the last statement in the forward process helps you choose the right key question and answer.

2.3 (c) is incorrect because it uses the specific notation given in the problem.

2.5 (a) is correct because it asks, abstractly, how the statement B can be proven true. (b) and (c) are not valid because they use the specific notation in the problem. Statement (d) is an incorrect question for this problem.

2.7 a. How can I show that
- two lines are parallel?
- two lines do not intersect?
- two lines tangent to a circle are parallel?
- two tangent lines passing through the endpoints of the diameter of a circle are parallel?

b. How can I show that
- a function is continuous?
- the sum of two functions is continuous?
- the sum of two continuous functions is continuous?

c. How can I show that
- an integer is even?
- an integer is not odd?
- an integer squared is even?

d. How can I show that
- the solution to a second-degree polynomial is a specific integer?
- a quadratic equation has a particular solution?
- two quadratic equations share a common root?

2.9 a. Show that
- their difference equals zero.

one is $\leq$ the other and vice versa.
their ratio is one.
they are both equal to a third number.

b. Show that
their corresponding side-angle-sides are equal.
their corresponding angle-side-angles are equal.
their corresponding side-side-sides are equal.
they are both congruent to a third triangle.

c. Show that
they lie in the same plane and do not intersect.
they are both perpendicular to a third line and lie in the same plane.
they have equal slopes.
their corresponding equations are identical or have no common solution.
they are each parallel to a third line in the plane.

2.11 a.
1. How can I show that the solution to a quadratic equation is positive?
2. Show that the quadratic formula gives a positive solution.
3. Show that the solution $\frac{-b}{2a}$ is positive.

b.
1. How can I show that a triangle is equilateral?
2. Show that three sides have equal length (or show that three angles are equal).
3. Show that $\overline{RT} = \overline{ST} = \overline{SR}$ [or show that angle (R) = angle (S) = angle (T)].

2.13 a. $(x-2)(x-1) < 0$.
$x(x-3) < -2$.
$-x^2 + 3x - 2 > 0$.

b. $\frac{x}{z} = \frac{1}{\sqrt{2}}$.
Angle (X) is a 45 degree angle.
$\cos(X) = \frac{1}{\sqrt{2}}$.

c. The circle has its center at $(3, 2)$.
The circle has a radius of 5.

The circle crosses the y axis at $(0,6)$ and $(0,-2)$.
$x^2 - 6x + 9 + y^2 - 4y + 4 = 25$.

2.15 (d) is not valid because "$x \neq 5$" is not stated in the hypothesis, and so, if $x = 5$, it will not be possible to divide by $x - 5$.

2.17 a. ***Analysis of proof.*** A key question associated with the conclusion is "How can I show that a real number (namely, x) is 0?" To show that $x = 0$, it will be established that both

B1: $x \leq 0$ and $x \geq 0$.

Working forward from the hypothesis immediately establishes that

A1: $x \geq 0$.

To see that $x \leq 0$, it will be shown that

B2: $x = -y$ and $-y \leq 0$.

Both of these statements follow by working forward from the hypothesis that

A2: $x + y = 0$ (so $x = -y$) and

A3: $y \geq 0$ (so $-y \leq 0$).

It remains only to show that

B3: $y = 0$,

which follows by working forward from the fact that

A4: $x = 0$

and the hypothesis that

A5: $x + y = 0$,

so,

A6: $0 = x + y = 0 + y = y$.

b. ***Proof.*** To see that both $x = 0$ and $y = 0$, it will first be shown that $x \geq 0$ (which is given in the hypothesis) and $x \leq 0$. The latter is accomplished by showing that $x = -y$ and that $-y \leq 0$. To see that $x = -y$, observe that the hypothesis states that $x + y = 0$. Similarly, $-y \leq 0$ because the hypothesis states that $y \geq 0$. Thus, $x = 0$. Finally, to see that $y = 0$, one can use the fact that $x = 0$ and the hypothesis that $x + y = 0$ to reach the desired conclusion. ∎

2.19 a. The number to the left of each line in Figure S.1 indicates which rule is being used.

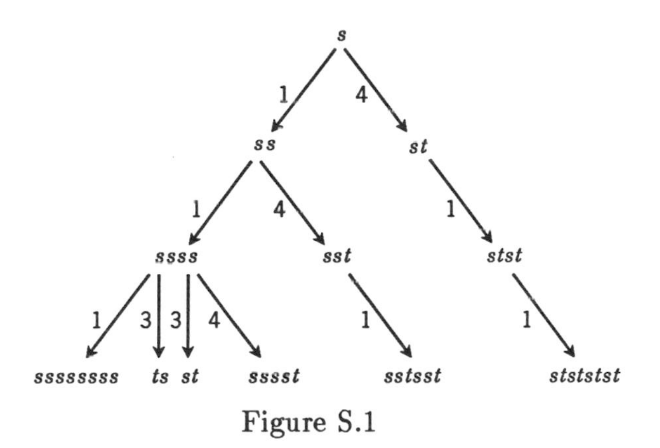

Figure S.1

b. The number to the left of each line in Figure S.2 indicates which rule is being used.

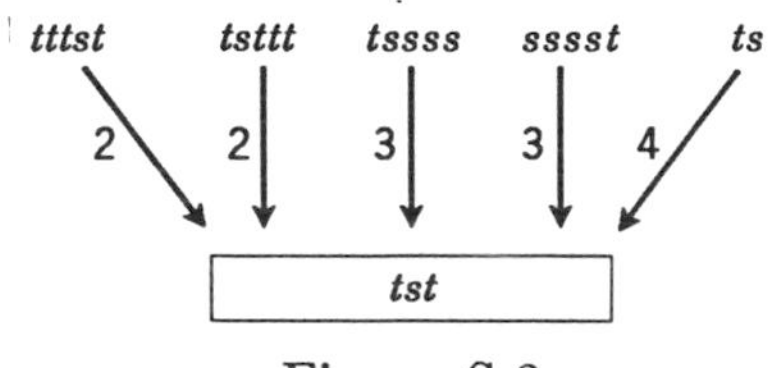

Figure S.2

c.

A :	s	given
A1 :	ss	rule 1
A2 :	$ssss$	rule 1
B1 :	$sssst$	rule 4
B :	tst	rule 3

d.

A :	s	given
A1 :	tst	from part (c)
A2 :	$tsttst$	rule 1
A3 :	$tsst$	rule 2
A4 :	$tssttsst$	rule 1
B1 :	$tssssst$	rule 2
B :	$ttst$	rule 3

2.21 ***Analysis of proof.*** In this problem one has:

A: The right triangle XYZ is isosceles.

B: The area of triangle XYZ is $\frac{z^2}{4}$.

A key question for B is "How can I show that the area of a triangle is equal to a particular value?" One answer is to use the formula for computing the area of a triangle to show that

B1: $\frac{z^2}{4} = \frac{xy}{2}$.

Working forward from the hypothesis that triangle XYZ is isosceles, one has:

A1: $x = y$, so

A2: $x - y = 0$.

Since XYZ is a right triangle, from the Pythagorean theorem,

A3: $z^2 = x^2 + y^2$.

Squaring both sides of the equation in $A2$ and performing some algebraic manipulations yields

A4: $(x - y)^2 = 0$.

A5: $x^2 - 2xy + y^2 = 0$.

A6: $x^2 + y^2 = 2xy$.

Substituting $A3$ into $A6$ yields

A7: $z^2 = 2xy$.

Dividing both sides by 4 finally yields the desired result:

A8: $\frac{z^2}{4} = \frac{xy}{2}$.

Proof. From the hypothesis one knows that $x = y$, or equivalently, that $x - y = 0$. Performing some algebraic manipulations yields $x^2 + y^2 = 2xy$. By the Pythagorean theorem, $z^2 = x^2 + y^2$ and on substituting z^2 for $x^2 + y^2$, one obtains $z^2 = 2xy$, or, $\frac{z^2}{4} = \frac{xy}{2}$. From the formula for the area of a right triangle, the area of $XYZ = \frac{xy}{2}$. Hence $\frac{z^2}{4}$ is the area of the triangle. ∎

2.23 ***Analysis of proof.*** A key question associated with the conclusion is "How can I show that a triangle is equilateral?" One answer is to show that all three sides have equal length, specifically,

B1: $\overline{RS} = \overline{ST} = \overline{RT}$.

To see that $\overline{RS} = \overline{ST}$, one can work forward from the hypothesis to establish that

B2: triangle RSU is congruent to triangle SUT.

Specifically, from the hypothesis, SU is a perpendicular bisector of RT, so

A1: $\overline{RU} = \overline{UT}$.

In addition,

A2: angle RUS = angle SUT = 90 degrees.

A3: $\overline{SU} = \overline{SU}$.

Thus the side-angle-side theorem states that the two triangles are congruent and so $B2$ has been established. It remains (from $B1$) to show that

B3: $\overline{RS} = \overline{RT}$.

Working forward from the hypothesis you can obtain this because

$$\text{A4:} \quad \overline{RS} = 2\overline{RU} = \overline{RU} + \overline{UT} = \overline{RT}.$$

Proof. To see that triangle RST is equilateral, it will be shown that $\overline{RS} = \overline{ST} = \overline{RT}$. To that end, the hypothesis that SU is a perpendicular bisector of RT ensures (by the side-angle-side theorem) that triangle RSU is congruent to triangle SUT. Hence, $\overline{RS} = \overline{ST}$. To see that $\overline{RS} = \overline{RT}$, by the hypothesis, one can conclude that $\overline{RS} = 2\overline{RU} = \overline{RU} + \overline{UT} = \overline{RT}$. ∎

SOLUTIONS TO CHAPTER 3

3.1
a. Key Qn: How can I show that an integer is odd?
Abs. Ans: Show that the integer equals two times some integer plus one.
Spec. Ans: Show that $n^2 = 2k + 1$ for some integer k.

b. Key Qn: How can I show that a real number is rational?
Abs. Ans: Show that the real number is equal to the ratio of two integers in which the denominator is not equal to zero.
Spec. Ans: Show that $\frac{s}{t} = \frac{p}{q}$ where p and q are integers and $q \neq 0$.

c. Key Qn: How can I show that two pairs of real numbers are equal?
Abs. Ans: Show that the first and second elements of one pair of real numbers are equal to the corresponding elements of the other pair.
Spec. Ans: Show that $x_1 = x_2$ and $y_1 = y_2$.

d. Key Qn: How can I show that an integer is prime?
Abs. Ans: Show that the integer is greater than 1 and can be divided only by itself and one.

Spec. Ans: Show that $n > 1$ and, if p is an integer that divides n, then $p = 1$ or $p = n$.

e. Key Qn: How can I show that an integer divides another integer?
Abs. Ans: Show that the second integer equals the product of the first integer with another integer.
Spec. Ans: Show that the following expression is true for some integer k: $(n-1)^3 + n^3 + (n+1)^3 = 9k$.

3.3 (A is the hypothesis; $A1$ is obtained by working forward step.)

a. **A :** n is an odd integer.
A1: $n = 2k + 1$ where k is an integer.

b. **A :** s and t are rational numbers with $t \neq 0$.
A1: $s = \frac{p}{q}$ where p and q are integers with $q \neq 0$. Also $t = \frac{a}{b}$ where $a \neq 0$ and $b \neq 0$ are integers.

c. **A :** Triangle RST is equilateral (see Figure S.3).
A1: $\overline{RS} = \overline{ST} = \overline{RT}$, and angle (R) = angle (S) = angle (T).

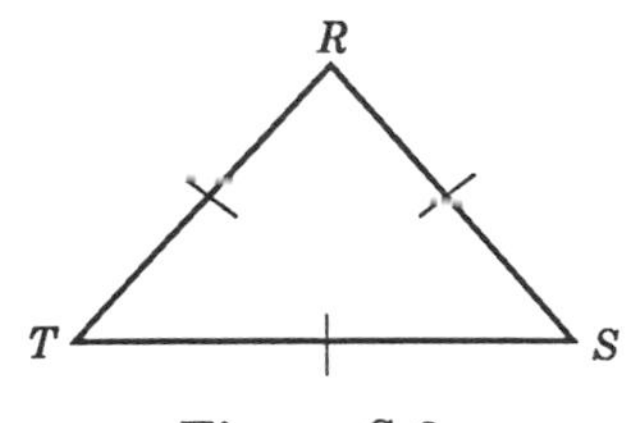

Figure S.3

d. **A :** $\sin(X) = \cos(X)$.
A1: $\frac{x}{z} = \frac{y}{z}$ (or, $x = y$).

e. **A :** a, b, c are integers for which a divides b and b divides c.
A1: $b = pa$ and $c = qb$ where p and q are two integers.

3.5 (T = true, F = false)

a. Truth table for the converse of "A implies B."

A	B	"A Implies B"	"B Implies A"
T	T	T	T
T	F	F	T
F	T	T	F
F	F	T	T

b. Truth table for the inverse of "A implies B."

A	B	NOT A	NOT B	"NOT A Implies NOT B"
T	T	F	F	T
T	F	F	T	T
F	T	T	F	F
F	F	T	T	T

The converse and inverse of "A implies B" are equivalent. Both are false if and only if A is false and B is true.

c. Truth table for "A OR B."

A	B	"A OR B"
T	T	T
T	F	T
F	T	T
F	F	F

d. Truth table for "*A* AND *B*."

A	*B*	"*A* AND *B*"
T	T	T
T	F	F
F	T	F
F	F	F

e. Truth table for "*A* AND NOT *B*"

A	*B*	NOT *B*	"*A* AND NOT *B*"
T	T	F	F
T	F	T	T
F	T	F	F
F	F	T	F

f. Truth table for "(NOT *A*) OR *B*"

A	NOT *A*	*B*	(NOT *A*) OR *B*	"*A* Implies *B*"
T	F	T	T	T
T	F	F	F	F
F	T	T	T	T
F	T	F	T	T

"*A* implies *B*" and "(NOT *A*) OR *B*" are equivalent. Both are false when *A* is true and *B* is false.

3.7 a. Converse: if n is an even integer, then n^2 is even.
Inverse: if n is an integer for which n^2 is odd, then n is odd.
Contrapositive: if n is an odd integer, then n^2 is odd.

b. Converse: if r is not rational, then r is a real number such that $r^2 = 2$.
Inverse: if r is a real number such that $r^2 \neq 2$, then r is rational.
Contrapositive: if r is rational, then r is a real number such that $r^2 \neq 2$.

c. Converse: if the quadrilateral $ABCD$ is a rectangle, then the quadrilateral $ABCD$ is a parallelogram with one right angle.
Inverse: if the quadrilateral $ABCD$ is not a parallelogram with one right angle, then the quadrilateral $ABCD$ is not a rectangle.
Contrapositive: if the quadrilateral $ABCD$ is not a rectangle, then $ABCD$ is not a parallelogram with one right angle.

d. Converse: if $t = \frac{\pi}{4}$ then t is an angle for which $\sin(t) = \cos(t)$.
Inverse: if t is an angle for which $\sin(t) \neq \cos(t)$, then $t \neq \frac{\pi}{4}$.
Contrapositive: if $t \neq \frac{\pi}{4}$ then t is an angle for which $\sin(t) \neq \cos(t)$.

3.9 ***Analysis of proof.*** The forward–backward method gives rise to the key question "How can I show an integer (namely, n^2) is odd?" The definition for an odd integer is used to answer the key question, which means you have to show that

B1: $n^2 = 2k + 1$ for some integer k.

We now turn toward the hypothesis and work forward to reach the desired conclusion.

Since n is an odd integer, by definition,

A1: $n = 2m + 1$ for some integer m.

Therefore,

A2: $n^2 = (2m + 1)^2$, so

A3: $n^2 = 4m^2 + 4m + 1$, so

A4: $n^2 = 2(2m^2 + 2m) + 1$.

Thus the desired value for k in $B1$ is $(2m^2 + 2m)$, and since m is an integer, $(2m^2 + 2m)$ is an integer. Hence it has been shown that n^2 can be expressed as 2 times some integer plus 1, which completes the proof.

Proof. Since n is an odd integer, there is an integer m for which $n = 2m + 1$, and therefore,

$$\begin{aligned} n^2 &= (2m + 1)^2 \\ &= 4m^2 + 4m + 1 \\ &= 2(2m^2 + 2m) + 1. \end{aligned}$$

Hence n^2 is an odd integer. ∎

3.11 ***Analysis of proof.*** Proceeding by the forward–backward method one is led to the key question "How can I show an integer (namely, mn) is odd?" Using the definition of an odd integer, this question can be answered by showing that

B1: $mn = 2k + 1$ for some integer k.

We now turn to the forward process to find out which integer.

Since m is an odd integer and n is an odd integer, one has, by definition,

A1: $m = 2b + 1$ for some integer b.

A2: $n = 2c + 1$ for some integer c.

Therefore,

A3: $mn = (2b+1)(2c+1)$, so

A4: $mn = 4bc + 2b + 2c + 1$, so

A5: $mn = 2(b + c + 2bc) + 1$.

Thus it has been shown that mn can be written as $2k+1$, where $k = (b + c + 2bc)$, and this completes the proof.

Proof. Since m and n are odd, there are integers b and c such that $m = 2b + 1$ and $n = 2c + 1$. So mn is odd since

$$\begin{aligned} mn &= (2b+1)(2c+1) \\ &= 2(b+c+2bc)+1. \quad \blacksquare \end{aligned}$$

3.13 ***Analysis of proof.*** Using the forward–backward method one is led to the key question "How can I show that one statement (namely, A) implies another statement (namely, C)?" According to Table 1 the answer is to assume that the statement to the left of the word "implies" is true, and then reach the conclusion that the statement to the right of the word "implies" is true. In this case, you assume

A1: A is true,

and try to reach the conclusion that

B1: C is true.

Working forward from the information given in the hypothesis, since "A implies B" is true and A is true, by Table 1 it must be that

A2: B is true.

Since B is true, and "B implies C" is true, it must also be that

A3: C is true.

Hence the proof is complete.

Proof. To conclude that "A implies C" is true, assume that A is true. By the hypothesis, "A implies B" is true, so B must be true. Finally, since "B implies C" is true, it must be that C is true, thus completing the proof. ∎

3.15 a. Six proofs are required to show that A is equivalent to each of the three alternatives. The six proofs are: $A \Rightarrow B$, $B \Rightarrow A$; $A \Rightarrow C$, $C \Rightarrow A$; $A \Rightarrow D$, and $D \Rightarrow A$.

b. Only four proofs are required: $A \Rightarrow B$, $B \Rightarrow C$, $C \Rightarrow D$, and $D \Rightarrow A$.

c. If the four statements in part (b) above are true, then you can show that A is equivalent to any of the alternatives by using Exercise 3.13. For instance, to show that A is equivalent to D, one already knows that "D implies A." By Exercise 3.13, since "A implies B," "B implies C," and "C implies D," one has "A implies D."

3.17 a. ***Analysis of proof.*** The forward–backward method gives rise to the key question "How can I show that a triangle is isosceles?" Using the definition of an isosceles triangle, you must show that two of its sides are equal which, in this case, means you must show that

B1: $u = v$.

Working forward from the hypothesis, you know that

A1: $\sin(U) = \sqrt{\frac{u}{2v}}$.

By the definition of sine, $\sin(U) = \frac{u}{w}$, so

A2: $\sqrt{\frac{u}{2v}} = \frac{u}{w}$,

and by algebraic manipulations, one obtains that

A3: $w^2 = 2uv$.

Furthermore, by the Pythagorean theorem,

A4: $u^2 + v^2 = w^2$.

Substituting for w^2 from $A3$ into $A4$ one has that

A5: $u^2 + v^2 = 2uv$, or,

A6: $u^2 - 2uv + v^2 = 0$.

On factoring $A6$, and then taking the square root of both sides of the equality, it follows that

A7: $u - v = 0$

and so $u = v$, completing the proof.

Proof. Since $\sin(U) = \sqrt{u/2v}$ and also $\sin(U) = u/w$, $\sqrt{u/2v} = u/w$, or, $w^2 = 2uv$. Now from the Pythagorean theorem, $w^2 = u^2 + v^2$, and on substituting $2uv$ for w^2, and then performing some algebraic manipulations, one has $u = v$. ∎

b. ***Analysis of proof.*** In order to verify the hypothesis of Example 1 for the current triangle UVW, it is necessary to match up the notation. Specifically, $x = u$, $y = v$, and $z = w$. Then it must be shown that

B1: $\frac{uv}{2} = \frac{w^2}{4}$.

Working forward from the current hypothesis that $\sin(U) = \sqrt{u/2v}$, and since $\sin(U) = u/w$, one has

A1: $\sqrt{u/2v} = u/w$, or,

A2: $u/2v = u^2/w^2$, or,

A3: $w^2 = 2uv$.

On dividing both sides of the equality in $A3$ by 4 yields precisely $B1$, thus completing the proof. (Observe also that triangle UVW is a right triangle.)

Proof. By the hypothesis, $\sin(U) = \sqrt{u/2v}$, and from the definition of sine, $\sin(U) = u/w$, thus $\sqrt{u/2v} = u/w$. By applying algebraic manipulations, one obtains $uv/2 = w^2/4$. Hence, the hypothesis of Example 1 holds for the current right triangle UVW, and consequently the triangle is isosceles. ∎

c. ***Analysis of proof.*** To verify the hypothesis of Example 3 for the current triangle UVW, it is necessary to match up the notation. Specifically, $r = u$, $s = v$, and $t = w$. Thus it must be shown that

B1: $w = \sqrt{2uv}$.

But, as in the proof of part (a),

A1: $\sqrt{u/2v} = u/w$, so

A2: $u/2v = u^2/w^2$, or,

A3: $w^2 = 2uv$.

On taking the positive square root of both sides of the equality in $A3$, one obtains precisely $B1$, thus completing the proof. (Observe also that triangle UVW is a right triangle.)

***Proof*.** By the hypothesis, $\sin(U) = \sqrt{u/2v}$ and from the definition of sine, $\sin(U) = u/w$, thus one has $\sqrt{u/2v} = u/w$. By algebraic manipulations one obtains $w = \sqrt{2uv}$. Hence the hypothesis of Example 3 holds for the current right triangle UVW, and consequently the triangle must be isosceles. ∎

SOLUTIONS TO CHAPTER 4

4.1

	Object	Certain property	Something happens
(a)	mountain in Himalayas	over 20,000 feet	taller than every other mountain in the world
(b)	integer x	none	$x^2 - \frac{5x}{2} + \frac{3}{2} = 0$
(c)	line ℓ'	through P not on ℓ	ℓ' is parallel to ℓ
(d)	angle t	$0 < t < \frac{\pi}{2}$	$\sin(t) = \cos(t)$
(e)	rational numbers r, s	between x and y	$\lvert r - s\rvert < 0.001$

4.3 a. A triangle XYZ is isosceles if $\exists$ two sides $\ni$ their lengths are equal.

b. Given an angle t, $\exists$ an angle t' $\ni$ the tangent of t' is greater than the tangent of t.

c. At a party of n people, $\exists$ at least two people $\ni$ they have the same number of friends.

d. For a polynomial of degree n, say $p(x)$, $\exists$ exactly n complex roots, $r_1, \ldots, r_n \ni p(r_1) = \ldots = p(r_n) = 0$.

4.5 ***Analysis of proof.*** The appearance of the key words "there is" in the conclusion of both parts suggests using the construction method to find values of x such that $x^2 - \frac{5x}{2} + \frac{3}{2} = 0$. From the quadratic formula, the values of x are:

$$
\begin{aligned}
x &= \left(\frac{\frac{5}{2} \pm \sqrt{\frac{25}{4} - \frac{12}{2}}}{2}\right) \\
&= \frac{\left(\frac{5}{2} \pm \frac{1}{2}\right)}{2} \\
&= 1 \text{ or } \tfrac{3}{2}.
\end{aligned}
$$

On substituting these values of x into $x^2 - \frac{5x}{2} + \frac{3}{2} = 0$, the quadratic equation is satisfied.

a. ***Proof.*** The quadratic formula for $x^2 - \frac{5x}{2} + \frac{3}{2} = 0$ yields $x = 1$ or $x = \frac{3}{2}$. Thus, there exists an integer, namely, $x = 1$, such that $x^2 - \frac{5x}{2} + \frac{3}{2} = 0$. The integer is unique. ∎

b. ***Proof.*** The quadratic formula for $x^2 - \frac{5x}{2} + \frac{3}{2} = 0$ yields $x = 1$ or $x = \frac{3}{2}$. Thus there exists a real number x such that $x^2 - \frac{5x}{2} + \frac{3}{2} = 0$. The real number x is not unique. ∎

4.7 ***Analysis of proof.*** The forward–backward method gives rise to the key question "How can I show that an integer (namely, a) divides another integer (namely, c)?" By the definition, one answer is to show that

B1: there is an integer k such that $c = ak$.

The appearance of the quantifier "there is" in $B1$ suggests turning to the forward process to construct the desired k.

From the hypothesis that $a|b$ and $b|c$, and by definition,

> **A1:** there are integers p and q such that $b = ap$ and $c = bq$.

Therefore, it follows that

> **A2:** $c = bq = (ap)q = a(pq)$,

and the desired integer k is $k = pq$.

Proof. Since $a|b$ and $b|c$, by definition, there are integers p and q for which $b = ap$ and $c = bq$. But then it follows that $c = bq = (ap)q = a(pq)$, and so $a|c$. ∎

4.9 ***Analysis of proof.*** The forward–backward method gives rise to the key question "How can I show that a real number (namely, $\frac{s}{t}$) is rational?" By the definition, one answer is to show that

> **B1:** there are integers p and q with $q \neq 0$ such that $\frac{s}{t} = \frac{p}{q}$.

The appearance of the quantifier "there are" suggests turning to the forward process to construct the desired p and q.

From the hypothesis that s and t are rational numbers, and by the definition,

> **A1:** there are integers a, b, c, and d with $b \neq 0$ and $d \neq 0$ such that $s = \frac{a}{b}$ and $t = \frac{c}{d}$.

Since $t \neq 0$, $c \neq 0$, and thus $bc \neq 0$. Hence,

> **A2:** $\frac{s}{t} = \frac{\left(\frac{a}{b}\right)}{\left(\frac{c}{d}\right)} = \frac{(ad)}{(bc)}$.

So the desired integers p and q are $p = ad$ and $q = bc$. Observe that since $b \neq 0$ and $c \neq 0$, $q \neq 0$; also, $\frac{s}{t} = \frac{p}{q}$.

Proof. Since s and t are rational, there are integers a, b, c, and d with $b \neq 0$ and $d \neq 0$ such that $s = \frac{a}{b}$ and $t = \frac{c}{d}$. Since $t \neq 0$, $c \neq 0$. Constructing $p = ad$ and $q = bc$, and noting that $q \neq 0$, one has $\frac{s}{t} = \left(\frac{a}{b}\right)/\left(\frac{c}{d}\right) = \frac{ad}{bc} = \frac{p}{q}$, and hence $\frac{s}{t}$ is rational. ∎

SOLUTIONS TO CHAPTER 5

5.1

a. Object: real number x.
Certain property: none.
Something happens: $f(x) \leq f(x^*)$.

b. Object: element x.
Certain property: x in S.
Something happens: $g(x) \geq f(x)$.

c. Object: element x.
Certain property: x in S.
Something happens: $x \leq u$.

d. Object: elements x and y, and real numbers t.
Certain property: x and y in C, and $0 \leq t \leq 1$.
Something happens: $tx + (1-t)y$ is an element of C.

e. Object: real numbers x, y, and t.
Certain property: $0 \leq t \leq 1$.
Something happens: $f(tx + (1-t)y) \leq tf(x) +$ $(1-t)f(y)$.

5.3

a. Let x' be a real number. It will be shown that $f(x') \leq f(x^*)$.

b. Let x' be an element in S. It will be shown that $g(x') \geq f(x')$.

c. Let x' be an element in S. It will be shown that $x' \leq u$.

d. Let x', y' be elements in C, and let t' be a real number between 0 and 1. It will be shown that $t'x' + (1-t')y'$ is an element of C.

e. Let x', y', t' be real numbers such that $0 \leq t' \leq 1$. It will be shown that $f(t'x' + (1 - t')y') \leq t'f(x') + (1 - t')f(y')$.

5.5 When using the choose method to show that "for all objects with a certain property, something happens," you would choose one particular object that does have the certain property. You would then work forward from the certain property to reach the conclusion that the something happens. This is precisely the same as using the forward–backward method to show that "if X is an object with the certain property, then the something happens," whereby you would work forward from the fact that X is an object with the certain property, and backward from the something that happens. Thus, a statement containing the quantifier "for all" can be converted into an equivalent statement having the form "if ... then ...."

5.7
a. $\exists$ a mountain $\ni$ $\forall$ other mountains, this one is taller than the others.
b. $\forall$ angle t, $\sin(2t) = 2\sin(t)\cos(t)$.
c. $\forall$ nonnegative real numbers p and q, $\sqrt{pq} \leq (p+q)/2$.
d. $\forall$ real numbers x and y with $x < y$, $\exists$ a rational number $r \ni x < r < y$.

5.9 ***Analysis of proof.*** Since the conclusion contains the key words "for all," the choose method is used to choose

A1: real numbers x and y with $x < y$,

for which it must be shown that

B1: $f(x) < f(y)$, i.e.,

B2: $mx + b < my + b$.

Work forward from the hypothesis that $m > 0$ to multiply both sides of the inequality $x < y$ in $A1$ by m yielding

A2: $mx < my$.

Adding b to both sides gives precisely $B2$.

5.11 ***Analysis of proof.*** The forward–backward method gives rise to the key question "How can I show that a set (namely, T) is a subset of another set (namely, S)?" The definition provides the answer that one must show that

B1: for all t in T, t is in S.

The appearance of the quantifier "for all" in the backward process suggests using the choose method to choose

A1: an element t' in T

for which it must be shown that

B2: t' is in S.

This in turn is done by showing that t' satisfies the defining property of S, that is, that

B3: $(t')^2 - 3t' + 2 \leq 0$.

Working forward from the fact that t' is in T, one has:

A2: $1 \leq t' \leq 2$, so

A3: $(t' - 1) \geq 0$ and $(t' - 2) \leq 0$, so

A4: $(t')^2 - 3t' + 2 = (t' - 1)(t' - 2) \leq 0$.

This establishes $B3$ and completes the proof.

Proof. To show that $T \subseteq S$, let t' be an element of T. It will be shown that t' is in S. Since t' is in T, $1 \leq t' \leq 2$, so $(t' - 1) \geq 0$ and $(t' - 2) \leq 0$. Thus, t' is in S since

$$(t')^2 - 3t' + 2 = (t' - 1)(t' - 2) \leq 0. \quad \blacksquare$$

5.13 ***Analysis of proof.*** The forward–backward method gives rise to the key question "How can I show a function is convex?" Using the definition in Exercise 5.1(e), one must show that

B1: for all real numbers x and y, and for all t with $0 \le t \le 1$, $f(tx+(1-t)y) \le tf(x)+(1-t)f(y)$.

The appearance of the quantifier "for all," suggests using the choose method whereby one chooses

A1: real numbers x' and y', and a real number t' that satisfies $0 \le t' \le 1$

for which it must be shown that

B2: $f(t'x' + (1-t')y') \le t'f(x') + (1-t')f(y')$.

But by the hypothesis,

A2:
$$\begin{aligned} f(t'x' + (1-t')y') &= m(t'x' + (1-t')y') + b \\ &= mt'x' + my' - mt'y' + b \\ &= mt'x' + bt' + my' - mt'y' + b - bt' \\ &= t'(mx' + b) + (1-t')(my' + b) \\ &= t'f(x') + (1-t')f(y'). \end{aligned}$$

Thus the desired inequality holds.

Proof. To show that f is convex, it will be shown that for all real numbers x and y, and for all t satisfying $0 \le t \le 1$, $f(tx + (1-t)y) \le tf(x) + (1-t)f(y)$. Let x' and y' be real numbers, and let t' satisfy $0 \le t' \le 1$, then

$$\begin{aligned} f(t'x' + (1-t')y') &= m(t'x' + (1-t')y') + b \\ &= mt'x' + my' - mt'y' + b \\ &= t'(mx' + b) + (1-t')(my' + b) \\ &= t'f(x') + (1-t')f(y'). \end{aligned}$$

Thus the inequality holds. ∎

SOLUTIONS TO CHAPTER 6

6.1 a. Applicable.

b. Not applicable because the statement contains the quantifier "there is" instead of "for all."

c. Applicable.

d. Applicable.

e. Not applicable because in this statement, n is a real number, and induction is applicable only to integers.

6.3 a. The choose method is used when the following form appears in the statement B: "for every object with a certain property, something happens." Induction is used whenever the object is an integer and the certain property is that of being greater than some initial integer. Induction is used in such cases because it is often easier to show that the something happens for $n+1$ given that it happens for n, rather than to show that it happens for n, given the certain property, as would be done with the choose method.

b. It is not possible to use induction when the object is not an integer because showing that $P(n)$ implies $P(n+1)$ may "skip over" many values of the object. As a result, the statement will not have been proved for such values.

6.5 ***Proof.*** First it must be shown that $P(n)$ is true for $n = 1$. Replacing n by 1, it must be shown that $1(1!) = (1+1)!-1$. But this is clear since $1(1!) = 1 = (1+1)! - 1$.

Now assume that $P(n)$ is true and use that fact to show that $P(n+1)$ is true. So assume

$$P(n) : 1(1!) + \ldots + n(n!) = (n+1)! - 1$$

It must be shown that

$$P(n+1) : 1(1!) + \ldots + (n+1)(n+1)! = (n+2)! - 1.$$

Starting with the left side of $P(n+1)$ and using $P(n)$:

$$\begin{aligned}
1(1!) + \ldots + n(n!) + (n+1)(n+1)! \\
&= [1(1!) + \ldots + n(n!)] + (n+1)(n+1)! \\
&= [(n+1)! - 1] + (n+1)(n+1)! \\
&= (n+1)![1 + (n+1)] - 1 \\
&= (n+1)!(n+2) - 1 \\
&= (n+2)! - 1. \quad \blacksquare
\end{aligned}$$

6.7 ***Proof.*** First it is shown that $P(n)$ is true for $n = 5$. But $2^5 = 32$ and $5^2 = 25$, so $2^5 > 5^2$. Hence it is true for $n = 5$. Assuming that $P(n)$ is true, one must then prove that $P(n+1)$ is true. So assume

$$P(n) : 2^n > n^2.$$

It must be shown that

$$P(n+1) : 2^{n+1} > (n+1)^2.$$

Starting with the left side of $P(n+1)$, and using the fact that $P(n)$ is true, one has:

$$2^{n+1} = 2(2^n) > 2(n^2).$$

To obtain $P(n+1)$, it must still be shown that for $n > 5$, $2n^2 > (n+1)^2 = n^2 + 2n + 1$, or, by subtracting $n^2 + 2n - 1$ from both sides and factoring, that $(n-1)^2 > 2$. This last statement is true because, for $n > 5$, $(n-1)^2 \geq 4^2 = 16 > 2$. $\blacksquare$

6.9 ***Proof.*** The statement is true for a set consisting of one element, say x, because its subsets are $\{x\}$ and $\emptyset$, that is, there are $2^1 = 2$ subsets. Assume that for a set with n elements, the number of subsets is 2^n. It will be shown that for a set with $(n+1)$ elements, the number of subsets is 2^{n+1}. For a set S with $(n+1)$ elements one can construct all the subsets by listing first those subsets that use the first n elements, and then, to each such subset, one can add the last element of S. By the induction hypothesis, there are 2^n subsets using the first n elements, and an additional 2^n subsets are created by adding the last element of S to each of the subsets of n elements. Thus the total number of subsets of S is $2^n + 2^n = 2^{n+1}$, and the statement has been established for $(n+1)$. ∎

6.11 ***Proof.*** Let

$$S = 1 + 2 + \ldots + n.$$

Then

$$S = n + (n-1) + \ldots + 1,$$

and on adding the two equations one obtains

$$2S = n(n+1), \text{ i.e., } S = n(n+1)/2. \quad ∎$$

6.13 ***Proof.*** For $n = 1$, $n^3 - n = 1 - 1 = 0$, and six divides 0 since $0 = (6)(0)$. Assuming that the statement is true for $(n-1)$, one knows that 6 divides $(n-1)^3 - (n-1)$, or, $(n-1)^3 - (n-1) = 6k$ for some integer k. It must be shown that $n^3 - n = 6m$ for some integer m. To relate $P(n)$ to $P(n-1)$ note that

$$\begin{aligned}(n-1)^3 - (n-1) &= n^3 - 3n^2 + 3n - 1 - n + 1 \\ &= n^3 - 3n^2 + 2n \\ &= n^3 - n - (3n^2 - 3n),\end{aligned}$$

so

$$n^3 - n = [(n-1)^3 - (n-1)] + 3n^2 - 3n.$$

By the induction hypothesis, $[(n-1)^3 - (n-1)]$ can be divided by 6, so it remains to show that $3n^2 - 3n$ can also be divided by 6, or equivalently, that $n^2 - n$ can be divided by 2. However, since $n^2 - n = n(n-1) =$ the product of two consecutive integers, either n or $(n-1)$ must be even, and so indeed, $n(n-1)$ can be divided by 2. Consequently, it follows that $3n^2 - 3n = 6p$ for some integer p. Thus the desired value for the integer m is $k + p$, for then

$$\begin{aligned}n^3 - n &= [(n-1)^3 - (n-1)] + 3(n^2 - n) \\ &= 6k + 6p \\ &= 6(k+p). \quad ∎\end{aligned}$$

6.15 a. Verify that the statement is true for the initial value. Then, assuming that the statement is true for n, prove that it is true for $n - 1$.

b. Verify that the statement is true for some integer. Assuming that the statement is true for n, prove that it is true for $n + 1$ and that it is also true for $n - 1$.

c. Verify that the statement is true for $n = 1$. Assuming that the statement is true for $2n + 1$, prove that it is also true for $2n + 3$. Alternatively, you can assume that $P(n)$ is true and show that $P(n + 2)$ is true.

6.17 The mistake occurs in the last sentence where it states that "Then, since all the colored horses in this (second) group are brown, the uncolored horse must also be brown." How do you know that *there is* a colored horse in the second group? In fact, when the original group of $(n + 1)$ horses consists of exactly 2 horses, the second group of n horses will not contain a colored horse. The entire difficulty is caused by the fact that the statement should have been verified for the initial integer $n = 2$, not $n = 1$! This, of course, you will not be able to do. ∎

SOLUTIONS TO CHAPTER 7

7.1 Use the choose method when the quantifier "for all" appears in the backward process. Use specialization when the quantifier "for all" appears in the forward process. In other words, use the choose method when you want to show that "for all objects with a certain property, something happens." Use specialization when you know that "for all objects with a certain property, something happens."

7.3 a. m must be an integer ≥ 5 and, if it is, then $2^m > m^2$.

b. y must be an element of S with $|y| < 5$ and, if it is, then y is an element of the set T.

c. The quadrilateral $QRST$ is a rectangle whose area is equal to the square of the length of a diagonal and, if it is, then $QRST$ is a square.

d. Angle S of triangle RST must be strictly between 0 and $\frac{\pi}{4}$ and, if it is, then $\cos(S) > \sin(S)$.

7.5 ***Analysis of proof.*** The forward–backward method gives rise to the key question "How can I show that a set (namely, R) is a subset of another set (namely, T)?" One answer is by the definition, so one must show that

B1: for all r in R, r is in T.

The appearance of the quantifier "for all" in the backward process suggests using the choose method. So choose

A1: an element r' in R

for which it must be shown that

B2: r' is in T.

Turning to the forward process, the hypothesis says that R is a subset of S and S is a subset of T. By definition, this means, respectively, that

A2: for all r in R, r is in S, and

A3: for all s in S, s is in T.

Specializing $A2$ to r' (which is in R), one has that

A4: r' is in S.

Specializing $A3$ to r' (which is in S), one has that

A5: r' is in T,

which is $B2$, thus completing the proof.

Proof To show that R is a subset of T, we must show that for all r in R, r is in T. Let r' be in R. By hypothesis, R

is a subset of S, so r' is in S. Also, by hypothesis, S is a subset of T, so r' is in T. ∎

7.7 ***Analysis of proof.*** The forward–backward method gives rise to the key question "How can I show that a set (namely, S intersect T) is convex?" One answer is by the definition, whereby it must be shown that

B1: for all x and y in S intersect T, and for all $0 \le t \le 1$, $tx + (1-t)y$ is in S intersect T.

The appearance of the quantifier "for all" in the backward process suggests using the choose method to choose

A1: x' and y' in S intersect T, and t' with $0 \le t' \le 1$,

for which it must be shown that

B2: $t'x' + (1-t')y'$ is in S intersect T.

Working forward from the hypothesis and $A1$, $B2$ will be established by showing that

B3: $t'x' + (1-t')y'$ is in both S and T.

Specifically, from the hypothesis that S is convex, and by the definition, it follows that

A2: for all x and y in S, and for all $0 \le t \le 1$, $tx + (1-t)y$ is in S.

Specializing this statement to $x = x'$, $y = y'$, and $t = t'$ (noting from $A1$ that $0 \le t' \le 1$) yields that

A3: $t'x' + (1-t')y'$ is in S.

A similar argument shows that $t'x' + (1-t')y'$ is also in T, thus completing the proof.

Proof. To see that S intersect T is convex, let x' and y' be in S intersect T, and let $0 \le t' \le 1$. It will be established that $t'x' + (1-t')y'$ is in S intersect T. From the hypothesis that S is convex, one has that $t'x'+(1-t')y'$ is in S. Similarly, $t'x'+(1-t')y'$ is in T. Thus it follows that $t'x'+(1-t')y'$ is in S intersect T, and so S intersect T is convex. ∎

7.9 ***Analysis of proof.*** The appearance of the quantifier "for all" in the conclusion indicates that the choose method should be used to choose

A1: a real number $s' \geq 0$

for which it must be shown that

B1: the function $s'f$ is convex.

An associated key question is "How can I show that a function (namely, $s'f$) is convex?" Using the definition in Exercise 5.1(e), one answer is to show that

B2: for all real numbers x and y, and for all $0 \leq t \leq 1$, $s'f(tx+(1-t)y) \leq ts'f(x) + (1-t)s'f(y)$.

The appearance of the quantifier "for all" in the backward process suggests using the choose method to choose

A2: real numbers x' and y', and $0 \leq t' \leq 1$

for which it must be shown that

B3: $s'f(t'x'+(1-t')y') \leq t's'f(x')+(1-t')s'f(y')$.

The desired result is obtained by working forward from the hypothesis that f is a convex function. By the definition in Exercise 5.1(e), one knows that

A3: for all real numbers x and y, and for all $0 \leq t \leq 1$, $f(tx+(1-t)y) \leq tf(x)+(1-t)f(y)$.

Specializing this statement to $x = x'$, $y = y'$, and $t = t'$ (noting that $0 \leq t' \leq 1$) yields

A4: $f(t'x' + (1 - t')y') \leq t'f(x') + (1 - t')f(y')$.

The desired statement $B3$ can be obtained by multiplying both sides of the inequality in $A4$ by the nonnegative number s', thus completing the proof.

Proof. To show that $s'f$ is convex, let x' and y' be real numbers, and let $0 \leq t' \leq 1$. It will be shown that $s'f(t'x' + (1 - t')y') \leq t's'f(x') + (1 - t')s'f(y')$.

Since f is a convex function by hypothesis, it then follows from the definition that $f(t'x' + (1 - t')y') \leq t'f(x') + (1-t')f(y')$. The desired result is obtained by multiplying both sides of this inequality by the nonnegative number s'. ∎

7.11 ***Analysis of proof.*** The forward–backward method gives rise to the key question "How can I show that a set (namely, C) is convex?" Using the definition in Exercise 5.1(d), one answer is to show that

> **B1:** for all x and z in the set C, and for all $0 \leq t \leq 1$, $tx + (1 - t)z$ is in C.

The appearance of the quantifier "for all" in the backward process suggests using the choose method to choose

A1: x' and z' in C, and t' with $0 \leq t' \leq 1$,

for which it must be shown that

B2: $t'x' + (1 - t')z'$ is in C.

This in turn is done by showing that $t'x'+(1-t')z'$ satisfies the defining property of C, i.e.,

B3: $f(t'x' + (1 - t')z') \leq y$.

Turning to the forward process, by hypothesis, f is a convex function. So by the definition in Exercise 5.1(e), it must be that

A2: for all x and z, and for all $0 \leq t \leq 1$,
$f(tx + (1-t)z) \leq tf(x) + (1-t)f(z)$.

Recognizing the quantifier "for all" in the forward process, specialization will be used. Specifically, $A2$ can be specialized to $x = x'$, $z = z'$, and $t = t'$ (noting that $0 \leq t' \leq 1$), so

A3: $f(t'x' + (1-t')z') \leq t'f(x') + (1-t')f(z')$.

To obtain $B3$, use the fact that x' and z' are in the set C, that is, they satisfy the defining property of C, thus,

A4: $f(x') \leq y$ and

A5: $f(z') \leq y$.

Multiplying both sides of $A4$ by the nonnegative number t' and both sides of $A5$ by the nonnegative number $(1-t')$ yields

A6: $t'f(x') \leq t'y$

A7: $(1-t')f(z') \leq (1-t')y$.

So, from $A3$, $A6$, and $A7$,

A8:
$$\begin{aligned} f(t'x' + (1-t')z') &\leq t'f(x') + (1-t')f(z') \\ &\leq t'y + (1-t')y \\ &= y, \end{aligned}$$

and hence $B3$ is true, completing the proof.

Proof. To show that C is convex, let x' and z' be elements of C, and let t' satisfy $0 \leq t' \leq 1$. Hence $f(x') \leq y$ and $f(z') \leq y$. By the hypothesis, f is a convex function, so

$$\begin{aligned} f(t'x' + (1-t')z') &\leq t'f(x') + (1-t')f(z') \\ &\leq t'y + (1-t')y \\ &= y. \end{aligned}$$

and consequently, $t'x' + (1-t')z'$ is in C. ∎

SOLUTIONS TO CHAPTER 8

8.1 a. For the quantifier "there is":
Object: real number y.
Certain property: none.
Something happens: for every real number x, $f(x) \leq y$.
For the quantifier "for every":
Object: real number x.
Certain property: none.
Something happens: $f(x) \leq y$.

b. For the quantifier "for all":
Object: real number ε.
Certain property: $\varepsilon > 0$.
Something happens: $\exists x \in S \ni x > u - \varepsilon$.
For the quantifier "there is":
Object: element x.
Certain property: $x \in S$.
Something happens: $x > u - \varepsilon$.

c. For the first quantifier "for all":
Object: real number ε.
Certain property: $\varepsilon > 0$.
Something happens: there is a real number $\delta > 0$ such that, for all real numbers y with $|x - y| < \delta$, $|f(x) - f(y)| < \varepsilon$.
For the quantifier "there is":
Object: real number δ.
Certain property: $\delta > 0$.
Something happens: for all real numbers y with $|x - y| < \delta$, $|f(x) - f(y)| < \varepsilon$.
For the second quantifier "for all":
Object: real number y.
Certain property: $|x - y| < \delta$.
Something happens: $|f(x) - f(y)| < \varepsilon$.

d. For the first quantifier "for all":
Object: real number ε.

Certain property: $\varepsilon > 0$.
Something happens: $\exists$ an integer $k' \ni \forall$ integers k with $k > k'$, $|x_k - x| < \varepsilon$.

For the quantifier "there is":
Object: integer k'.
Certain property: none.
Something happens: $\forall$ integers k with $k > k'$, $|x_k - x| < \varepsilon$.

For the second quantifier "for all":
Object: integer k.
Certain property: $k > k'$.
Something happens: $|x_k - x| < \varepsilon$.

8.3 a. Both $S1$ and $S2$ are true. This is because when you apply the choose method to each statement, in either case you will choose real numbers x and y with $0 \leq x \leq 1$ and $0 \leq y \leq 2$ for which you can then show that $2x^2 + y^2 \leq 6$.

b. $S1$ and $S2$ are different—$S1$ is true and $S2$ is false. The choose method can be used to show that $S1$ is true. To see that $S2$ is false, consider $y = 1$ and $x = 2$. For these real numbers, $2x^2 + y^2 = 2(4) + 1 = 9 > 6$.

c. These two statements are the same when the properties P and Q do not depend on the objects X and Y. This was the case in part (a) above but not in part (b).

8.5 a. The construction method would be used first to construct a real number $M > 0$. The choose method would be used next to show that, for the value of M you constructed, it is true that for all elements x in the set T, $|x| \leq M$. In so doing you would choose an element x in the set T for which you must then show that $|x| \leq M$.

b. First the choose method is used to choose a real number $M > 0$ for which it must be shown that there is an element x in the set T such that $|x| > M$. To show this, the construction method is used next whereby you must construct an element x in the set T; you must then show that this element x satisfies $|x| > M$.

c. The choose method is used first to choose a real number $\varepsilon > 0$ for which it must be shown that there is a real number $\delta > 0$ such that for all real numbers y with $|x - y| < \delta$, $|f(x) - f(y)| < \varepsilon$. To show this, the construction method is used next to construct a real number $\delta > 0$. You must then show that this value of δ satisfies the property that for all real numbers y with $|x-y| < \delta$, $|f(x)-f(y)| < \varepsilon$. To do this, use the choose method next to choose a real number y with $|x-y| < \delta$, for which you must show that $|f(x) - f(y)| < \varepsilon$.

8.7 ***Analysis of proof.*** The key word "for every" in the conclusion suggests using the choose method to choose

A1: a real number $x' > 2$

for which it must be shown that

B1: there is a real number $y < 0$ such that $x' = 2y/(1 + y)$.

The key words "there is" in $B1$ suggest using the construction method to construct the desired y. Working backward from the fact that y must satisfy $x' = 2y/(1+y)$, it follows that y must be constructed so that

B2: $x' + x'y = 2y$, or

B3: $y(2 - x') = x'$, or

B4: $y = \frac{x'}{(2-x')}$.

To see that this is the correct value for y, it is easily seen that $x' = 2y/(1+y)$. However, it must also be shown that $y < 0$, which it is, since $x' > 2$.

Proof. Let $x' > 2$; therefore it is possible to construct $y = x'/(2 - x')$. Since $x' > 2$, $y < 0$. It is also easy to verify that $x' = 2y/(1 + y)$. ∎

8.9 ***Analysis of proof.*** The forward–backward method gives rise to the key question "How can I show that a number (namely, 1) is a least upper bound for a set (namely, S)?" Using the definition in Exercise 8.1(b), one answer is to show that

B1: 1 is an upper bound for S, and, for all $\varepsilon > 0$, there is an element x in S such that $x > 1 - \varepsilon$.

To show that the first part of $B1$ is true, one is led to the key question "How can I show that a number (namely, 1) is an upper bound for a set (namely, S)?" Again, the definition in Exercise 5.1(c) can be used to provide the answer that one must show that

B2: for all x in S, $x \leq 1$.

Recognizing the quantifier "for all" in the backward process, the choose method is used to choose

A1: an element x in S

for which it must be shown that

B3: $x \leq 1$.

To establish $B3$, you can make use of the fact that x is in S (see $A1$). To do so, it is important to observe that the set S can be written as

$$S = \{\text{real numbers } x : \text{ there is an integer } n \geq 2 \text{ with } x = 1 - \tfrac{1}{n}\}.$$

Since x is in S,

A2: there is an integer $n \geq 2$ for which $x = 1 - \frac{1}{n}$,

and since $n \geq 2$, $B3$ is true because

A3: $x = 1 - \frac{1}{n} \leq 1$.

Returning to $B1$, one must still show that

B4: for all $\varepsilon > 0$, there is an element x in S such that $x > 1 - \varepsilon$.

Again, the choose method is used to select

A4: an $\varepsilon > 0$

for which it must be shown that

B5: there is an element x in S such that $x > 1 - \varepsilon$.

Turning to the forward process, the desired x in S will be constructed by finding an integer $n \geq 2$ for which $1 - \frac{1}{n} > 1 - \varepsilon$, for then one can construct $x = 1 - \frac{1}{n}$. The desired n is any integer $> \frac{1}{\varepsilon}$, for then $1 - \frac{1}{n} > 1 - \varepsilon$, thus completing the proof.

Proof. To see that 1 is an upper bound for S, let x be in S. Hence there is an integer $n \geq 2$ such that $x = 1 - \frac{1}{n}$. So $x \leq 1$, and therefore 1 is an upper bound for S. To complete the proof, let $\varepsilon > 0$. An element x in S will be produced for which $x > 1 - \varepsilon$. To do so, let n be any integer greater than $\frac{1}{\varepsilon}$. Then $x = 1 - \frac{1}{n}$ satisfies $x > 1 - \varepsilon$. ∎

SOLUTIONS TO CHAPTER 9

9.1 a. Assume: ℓ, m, and n are three consecutive integers, and that 24 divides $\ell^2 + m^2 + n^2 + 1$.

b. Assume: Matrix M is not singular, and the rows of M are linearly dependent.

c. Assume f and g are two functions such that $g \geq f$, f is unbounded above, and g is not unbounded.

9.3 a. The number of primes is not finite.

b. The set of real numbers is not bounded.

c. The positive integer p cannot be divided by any positive integer other than 1 and p.

d. The lines ℓ and ℓ' do not intersect.

e. The real number x is not ≥ 5.

9.5 a. Use the construction method to construct an element s in S and show that element is also in T.

b. First use the choose method to choose an element s in S for which it must be shown that there is no element t in T such that $s > t$. To show the latter, use the contradiction method whereby you should assume that s is an element of S and there is an element t in T such that $s > t$. Then you must reach a contradiction.

c. First use the contradiction method whereby you should assume that there is a real number $M > 0$ such that for all elements x in S, $|x| < M$. To reach a contradiction, you will probably have to apply specialization to the statement "for all elements x in S, $|x| < M$."

9.7 ***Analysis of proof.*** To use the contradiction method, assume:

A: n is an integer for which n^2 is even.
NOT B: n is not even, i.e., n is odd.

Work forward from these assumptions using the definition of an odd integer to reach the contradiction that

B1: n^2 is odd.

Applying a definition to work forward from NOT B yields

A1: there exists an integer k such that $n = 2k + 1$.

Squaring both sides of the equality in $A1$ and performing simple algebraic manipulations leads to

A2: $n^2 = (2k+1)^2$, so

A3: $n^2 = 4k^2 + 4k + 1$, so

A4: $n^2 = 2(2k^2 + 2k) + 1$,

which says that $n^2 = 2p + 1$, where $p = 2k^2 + 2k$. Thus n^2 is odd, and this contradiction establishes the result.

Proof. Assume, to the contrary, that n is odd and n^2 is even. Hence, there is an integer k such that $n = 2k + 1$. Consequently,

$$\begin{aligned} n^2 &= (2k+1)^2 \\ &= 4k^2 + 4k + 1 \\ &= 2(2k^2 + 2k) + 1, \end{aligned}$$

and hence n^2 is odd, contradicting the initial assumption. ∎

9.9 ***Analysis of proof.*** Proceed by assuming that there is a chord of a circle that is longer than its diameter (see Figure S.4). Using this assumption and the properties of a circle, you must arrive at a contradiction.

Let AC be the chord of the circle that is longer than the diameter of the circle. Let AB be a diameter of the circle. This construction is valid since, by definition, a diameter is a line passing through the center terminating at the perimeter of the circle.

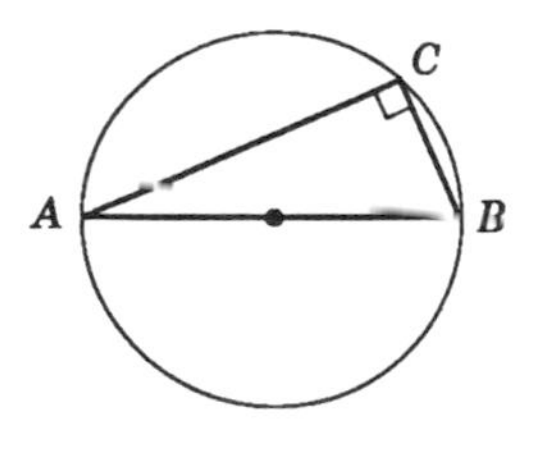

Figure S.4

It follows that angle ACB has 90 degrees (since it is an angle in a semi-circle). Hence the triangle ABC is a right triangle in which AB is the hypotenuse. Then the desired contradiction is that the hypotenuse of a right triangle is shorter than one side of the triangle, which is impossible.

Proof. Assume that there does exist a chord, say AC, of a circle that is longer than a diameter. Construct a diameter that has one of its ends coinciding with one end of the chord. Joining the other ends produces a right triangle in which the diameter AB is the hypotenuse. But then the hypotenuse would be shorter than one leg of the right triangle, which is a contradiction. ∎

9.11 ***Analysis of proof.*** The proof is done by contradiction, whereby it is assumed that a, b, c, x, y, and z are real numbers satisfying

A1: $az - 2by + cx = 0.$

A2: $ac - b^2 > 0.$

NOT B: $xz - y^2 > 0.$

A contradiction is reached by showing that the square of a number is less than 0, which cannot happen.

Working forward by adding b^2 to both sides of $A2$ and y^2 to both sides of NOT B, and multiplying the corresponding sides of these two inequalities (which are positive) yields

A3: $(ac)(xy) > b^2y^2.$

Working forward from $A1$ by adding $2by$ to both sides and then squaring the resulting equality yields

A4: $(az + cx)^2 = 4b^2y^2.$

Combining $A4$ with the inequality in $A3$ gives

A5: $(az + cx)^2 = 4b^2y^2 < 4(ac)(xy).$

Expanding the left side, collecting terms, and rewriting yields the desired contradiction that

A6: $(az - cx)^2 < 0.$

9.13 ***Analysis of proof.*** To use contradiction, assume that no two people have the same number of friends, that is, everybody has a different number of friends. Since there are n people, each of whom has a different number of friends, the people can be numbered in an increasing sequence according to the number of friends that person has. In other words,

person number 1 has	no	friends
person number 2 has	1	friend
person number 3 has	2	friends
...		
person number n has	$n-1$	friends.

By doing so, there is a contradiction that the last person will be friends with all the other $(n-1)$ people, including the first one, who has no friends!

Proof. Assume, to the contrary, that no two people have the same number of friends. The people at the party can be numbered in such a way that

person 1 has	0	friends
person 2 has	1	friend
...		
person n has	$n-1$	friends.

It then follows that the person with $(n-1)$ friends is a friend of the person who has no friends, a contradiction. ∎

9.15 ***Analysis of proof.*** By the contradiction method, it can be assumed that

A1: the number of primes is finite.

From $A1$, there will be a prime number that is larger than all the other prime numbers. So

A2: let n be the largest prime number.

Consider the number $n! + 1$, and let

A3: p be any prime number that divides $n! + 1$.

A contradiction will be reached by showing that

B1: $p \leq n$ and $p > n$.

Since n is the largest prime and p is prime, $p \leq n$. It remains to show that $p > n$. To do so, use the fact that p divides $n! + 1$ to show that

B2: $p \neq 2, p \neq 3, \ldots, p \neq n.$

To see that $B2$ is true, observe that when $n! + 1$ is divided by 2, there is a remainder of 1 since,

$$n! + 1 = [n(n-1)\ldots(2)(1)] + 1.$$

Similarly, when $n! + 1$ is divided by r, where $1 < r \leq n$, there is a remainder of 1. Indeed, one has that

$$\frac{(n!+1)}{r} = \frac{n!}{r} + \frac{1}{r}.$$

Hence $n! + 1$ has no prime factor between 1 and n. Therefore the prime factor p is greater than n, which contradicts the assumption that n is the largest prime number.

Proof. Assume, to the contrary, that there are a finite number of primes. Let n be the largest prime. Let p be any prime divisor of $n! + 1$. Since n is the largest prime, p must be less than or equal to n. But $n! + 1$ cannot be divided by any number between 1 and n. Hence $p > n$, which is a contradiction. ∎

SOLUTIONS TO CHAPTER 10

10.1

a. Work forward from: n is an odd integer.
Try to conclude: n^2 is an odd integer.

b. Work forward from: S is a subset of T, and T is bounded.
Try to conclude: S is bounded.

c. Work forward from: p and q are positive real numbers with $p = q$.
Try to conclude: $\sqrt{pq} = (p + q)/2$.

d. Work forward from: The rows of matrix M are linearly dependent.

Try to conclude: M is singular.

10.3 Statement (b) is a result of the forward process because you can assume that there is a real number t between 0 and $\frac{\pi}{4}$ such that $\sin(t) = r\cos(t)$. The answer in part (b) results on squaring both sides and replacing $\cos^2(t)$ with $(1 - \sin^2(t))$.

10.5 With the contrapositive method you work forward from NOT B and backward from NOT A. Therefore, the key question should be applied to NOT A, which is: the derivative of the function f at the point x is equal to 0.

a. Incorrect because the key question is being applied to NOT B. Also, the question uses symbols and notation from the specific problem.

b. Incorrect because the key question uses symbols and notation from the specific problem.

c. Incorrect because the key question is being applied to NOT B.

d. Correct.

10.7 *Analysis of proof.* With the contrapositive method, one can assume that

A1: there is an integer solution, say m, to the equation $n^2 + n - c = 0$. (NOT B)

It must be shown that

B1: c is not odd, i.e., that c is even. (NOT A)

But from $A1$, one can write that

A2: $c = m + m^2$.

To see that c is even, observe that $m + m^2 = m(m + 1)$ is the product of two consecutive integers and must thus be even.

Proof. Assume that there is an integer solution, say m, to the equation $n^2 + n - c = 0$. It will be shown that c is even. But $c = m + m^2 = m(m+1)$ is even because the product of two consecutive integers is even. ∎

10.9 ***Analysis of proof.*** By the contrapositive method, one can assume that

> **A1:** the quadrilateral $RSTU$ is not a rectangle. (NOT B)

It must be shown that

> **B1:** there is an obtuse angle. (NOT A)

The appearance of the quantifier "there is" in $B1$ suggests turning to the forward process to produce the obtuse angle.

Working forward from $A1$, one can conclude that

> **A2:** at least one angle of the quadrilateral is not 90 degrees, say it is angle R.

If angle R has more than 90 degrees, then it is the desired angle and the proof is complete. Otherwise,

> **A3:** angle R has less than 90 degrees.

This in turn means that

> **A4:** the remaining angles of the quadrilateral must add up to more than 270 degrees,

because the sum of all the angles in $RSTU$ is 360 degrees. Among these three angles that add up to more than 270 degrees, one of them must be greater than 90 degrees, and that is the desired obtuse angle.

Proof. Assume that the quadrilateral $RSTU$ is not a rectangle, and hence, one of its angles, say R, is not 90 degrees. An obtuse angle will be found. If angle R has more than 90 degrees, then it is the desired angle. Otherwise the remaining three angles add up to more than 270 degrees. Thus one of the remaining three angles is obtuse. ∎

SOLUTIONS TO CHAPTER 11

11.1

a. The real number x^* is not a maximizer of the function f if there is a real number x such that $f(x) > f(x^*)$.

b. Suppose that f and g are functions of one variable. Then g is not $\geq f$ on the set S of real numbers if there exists an element x in S such that $g(x) < f(x)$.

c. The real number u is not an upper bound for a set S of real numbers if there exists an x in S such that $x > u$.

d. The set C of real numbers is not a convex set if there exist elements x and y in C and there exists a real number t between 0 and 1 such that $tx + (1-t)y$ is not an element of C.

e. The function f of one variable is not convex if there exist real numbers x and y and t with $0 \leq t \leq 1$, such that $f(tx + (1-t)y) > tf(x) + (1-t)f(y)$.

11.3

a. There does not exist an element x in the set S such that x is not in T.

b. It is not true that for every angle t between 0 and $\frac{\pi}{2}$, $\sin(t) \neq \cos(t)$.

c. There does not exist an object with the certain property such that the something does not happen.

d. It is not true that for every object with the certain property, the something does not happen.

11.5

a. Work forward from NOT B.
Work backward from (NOT A) OR (NOT C).

b. Work forward from NOT B.
   Work backward from (NOT A) AND (NOT C).

c. Work forward from (mn is not divisible by 4) AND (n is divisible by 4)
   Work backward from (n is an odd integer) OR (m is an even integer).

11.7 ***Analysis of proof.*** When using the contradiction method, it can be assumed that

A1: $x \geq 0$, $y \geq 0$, $x + y = 0$, and

A2: either $x \neq 0$ or $y \neq 0$. (NOT B)

From $A2$, suppose first that

A3: $x \neq 0$.

Since $x \geq 0$ from $A1$, it must be that

A4: $x > 0$.

A contradiction to the fact that $y \geq 0$ will be reached by showing that

B1: $y < 0$.

Specifically, since $x + y = 0$ from $A1$,

A5: $y = -x$.

Since $-x < 0$ from $A4$, a contradiction has been reached. A similar argument can be used for the case where $y \neq 0$ (see $A2$).

Proof. Assume that $x \geq 0$, $y \geq 0$, $x + y = 0$, and that either $x \neq 0$ or $y \neq 0$. If $x \neq 0$, then $x > 0$ and $y = -x < 0$, but this contradicts the fact that $y \geq 0$. Similarly, if $y \neq 0$, then $y > 0$, and $x = -y < 0$, but this contradicts the fact that $x \geq 0$. ∎

SOLUTIONS TO CHAPTER 12

12.1 ***Analysis of proof.*** The direct uniqueness method will be used, whereby one must first construct the real number y. This was done in Exercise 8.7. It remains to show the uniqueness by assuming that y and z are real numbers with

A1: $y < 0.$

A2: $z < 0.$

A3: $x = \dfrac{2y}{(1+y)}.$

A4: $x = \dfrac{2z}{(1+z)}.$

Working forward via algebraic manipulations, it will be shown that

B1: $y = z.$

Specifically, combining $A3$ and $A4$ yields

A5: $x = \dfrac{2y}{(1+y)} = \dfrac{2z}{(1+z)}$

or, dividing both sides by 2 and clearing the denominators,

A6: $y + yz = z + yz.$

On subtracting yz from both sides one obtains the desired conclusion that $y = z$.

Proof. The existence of the real number y was established in Exercise 8.7. To show that y is unique, suppose that y and z satisfy $y < 0$, $z < 0$, $x = \frac{2y}{(1+y)}$, and also $x = \frac{2z}{(1+z)}$. But then $\frac{2y}{(1+y)} = \frac{2z}{(1+z)}$, and so $y + yz = z + yz$, or, $y = z$, as desired. ∎

12.3 ***Analysis of proof.*** According to the indirect uniqueness method, one must first construct a real number x for which $mx + b = 0$. But since the hypothesis states the $m \neq 0$, the desired x is $\frac{-b}{m}$, because then it follows that $mx + b = m\left(\frac{-b}{m}\right) + b = -b + b = 0$.

To establish the uniqueness by the indirect uniqueness method, suppose that x and y satisfy

A1: $mx + b = 0$.

A2: $my + b = 0$.

A3: $x \neq y$.

A contradiction to the hypothesis that $m \neq 0$ is reached by showing that

B1: $m = 0$.

Specifically, from $A1$ and $A2$,

A4: $mx + b = my + b$.

Subtracting the right side of the equality in $A4$ from the left side and rewriting yields

A5: $m(x - y) = 0$.

On dividing both sides of the equality in $A5$ by the nonzero number $x - y$ (see $A3$), it follows that $m = 0$. This contradiction establishes the uniqueness.

Proof. To construct the number x for which $mx + b = 0$, let $x = \frac{-b}{m}$ (since $m \neq 0$). Then $mx + b = m\left(\frac{-b}{m}\right) + b = 0$. Now suppose that $y \neq x$ and also satisfies $my + b = 0$. Then $mx + b = my + b$, and so $m(x - y) = 0$. But since $x - y \neq 0$, it must be that $m = 0$, and this contradicts the hypothesis that $m \neq 0$. ∎

12.5 ***Analysis of proof.*** The issue of existence must be addressed first. To construct the desired complex number

$(c + di)$ that satisfies $(a + bi)(c + di) = 1$, multiply the two terms using complex arithmetic to get $ac - bd = 1$ and $bc + ad = 0$. Solving these two equations for the two unknowns c and d in terms of a and b leads you to construct $c = a/(a^2 + b^2)$ and $d = -b/(a^2 + b^2)$ (noting that the denominator is not 0 since, by the hypothesis, at least one of a or b is not 0). To see that this construction is correct, note that

$$\textbf{A1:} \quad \begin{aligned} (a + bi)(c + di) &= (ac - bd) + (bc + ad)i \\ &= \left[\tfrac{(a^2+b^2)}{(a^2+b^2)}\right] + 0i \\ &= 1. \end{aligned}$$

To see the uniqueness, suppose that $e + fi$ is another complex number that also satisfies

$$\textbf{A2:} \quad (e + fi)(a + bi) = 1.$$

It will be shown that

$$\textbf{B1:} \quad (c + di) = (e + fi).$$

Working forward and multiplying both sides of the equality in $A1$ by $e + fi$ and using associativity, one obtains

$$\textbf{A3:} \quad (e + fi)[(a + bi)(c + di)] = (e + fi).$$

Since $(e + fi)(a + bi) = 1$ from $A2$, it follows from $A3$ that $c + di = e + fi$, and so $B1$ is true, completing the proof.

Proof. Since either $a \neq 0$ or $b \neq 0$, $a^2 + b^2 \neq 0$, and so it is possible to construct the complex number $c + di$ in which $c = a/(a^2 + b^2)$ and $d = -b/(a^2 + b^2)$, for then

$$(a + bi)(c + di) = ac - bd + (bc + ad)i = 1.$$

To see the uniqueness, suppose that $e + fi$ also satisfies $(a + bi)(e + fi) = 1$. By the rules of complex multiplication, it follows that

$$(e + fi)[(a + bi)(c + di)] = (e + fi)(1)$$

so $c + di = e + fi$ and the uniqueness is established. ∎

12.7 a. Advantage: You have three statements to work forward from: A, NOT C, and NOT D. With an either/or method, you would only have two statements to work forward from (A and NOT C).
Disadvantage: You cannot work backward because you do not know what the contradiction will be. With an either/or method you can work backward from the statement D.

b. To apply a proof by elimination to the statement "If A then C OR D OR E," you would assume that A is true, C is not true, and D is not true; you must conclude that E is true. (Alternatively, you can assume that A is true and that any two of the three statements C, D, and E are not true; you would then have to conclude that the remaining statement is true.)

c. To apply a proof by cases to the statement "If C OR D OR E, then B," you must do all three of the following proofs: (1) "If C then B," (2) "If D then B," and (3) "If E then B."

12.9 ***Analysis of proof.*** By the either/or method, assume that

A1: m and n are integers, and

A2: 4 divides n. (NOT D)

It must be shown that

B1: 4 divides mn. (C)

Working backward from $B1$, one is led to the key question "How can I show one integer (namely, 4) divides another integer (namely, mn)?" Using the definition leads to the answer that you must show that

B2: there is an integer k such that $mn = 4k$.

The appearance of the quantifier "there is" suggests using the construction method. Turning to the forward process, you must construct the value for k. Working forward from $A2$ by definition,

A3: there is an integer p such that $n = 4p$.

On multiplying both sides of the equality in $A3$ by m, one has

A4: $mn = 4mp$.

From $A4$ you can see that the desired value of k in $B2$ is mp, thus completing the proof. (This proof could also be done by assuming m and n are integers and that 4 does not divide mn; you would then have to show 4 does not divide n.)

Proof. Assume that m and n are integers, and that 4 divides n. It will be shown that 4 divides mn, or equivalently, that there is an integer k such that $mn = 4k$. But since 4 divides n, there is and integer p such that $n = 4p$. Thus, $mn = 4mp$, and so $k = mp$. ∎

12.11 a. If x is a real number that satisfies $x^3+3x^2-9x-27 \geq 0$, then $x \leq -3$ or $x \geq 3$.

b. ***Analysis of proof.*** According to the either/or method, you can assume that

A1: $x^3 + 3x^2 - 9x - 27 \geq 0$, and

A2: $x > -3$. (NOT C)

It must be shown that

B1: $x \geq 3$, or $x - 3 \geq 0$. (D)

By factoring $A1$, it follows that

A3: $x^3+3x^2-9x-27 = (x-3)(x+3)^2 \geq 0$.

From $A2$, since $x > -3$, $(x + 3)^2$ is strictly positive. Thus, dividing both sides of $A3$ by $(x+3)^2$ yields $B1$.

Proof. Assume that $x^3 + 3x^2 - 9x - 27 \geq 0$ and $x > -3$. Then it follows that

$$x^3 + 3x^2 - 9x - 27 = (x - 3)(x + 3)^2 \geq 0.$$

Since $x > -3$, $(x + 3)^2$ is positive, so $x - 3 \geq 0$, or equivalently, $x \geq 3$. ∎

c. ***Analysis of proof.*** According to this either/or method, you can assume that

A1: $x^3 + 3x^2 - 9x - 27 \geq 0$, and

A2: $x < 3$. (NOT D)

It must be shown that

B1: $x \leq -3$, or $x + 3 \leq 0$. (C)

By factoring $A1$, it follows that

A3: $x^3 + 3x^2 - 9x - 27 = (x - 3)(x + 3)^2 \geq 0.$

From $A2$, since $x < 3$, $(x - 3) < 0$. Thus, dividing both sides of $A3$ by $(x - 3)$ yields

A4: $(x + 3)^2 \leq 0.$

Since $(x + 3)^2$ is also ≥ 0, from $A4$, it must be that

A5: $(x + 3)^2 = 0$, so

A6: $x + 3 = 0$, or $x = -3$.

Thus $B1$ is true, completing the proof.

Proof. Assume that $x^3 + 3x^2 - 9x - 27 \geq 0$ and $x < 3$. Then it follows that

$$x^3 + 3x^2 - 9x - 27 = (x - 3)(x + 3)^2 \geq 0.$$

Since $x < 3$, $(x + 3)^2$ must be 0, so $x + 3 = 0$, or equivalently, $x = -3$. ∎

12.13 ***Analysis of proof.*** The appearance of the key words either/or in the hypothesis suggest proceeding with a proof by cases.

Case 1. Assume that

A1: $a|b$.

It must be shown that

B1: $a|(bc)$.

$B1$ gives rise to the key question "How can I show that an integer (namely, a) divides another integer (namely, bc)?" Applying the definition means you must show that

B2: there is an integer k such that $bc = ka$.

Recognizing the key words "there is" in $B2$, you should use the construction method to produce the desired integer k. Working forward from $A1$ by definition, you know that

A2: there is an integer p such that $b = pa$.

Multiplying both sides of the equality in $A2$ by c yields

A3: $bc = cpa$.

From $A3$, it is easy to see that the desired value for k is cp, thus completing this case.

Case 2. In this case, you should assume that

A1: $a|c$.

You must show that

B1: $a|(bc)$.

The remainder of the proof in this case is similar to that in Case 1 and will not be repeated.

Proof. Assume, without loss of generality, that $a|b$. By definition, there is an integer p such that $b = pa$. But then, $bc = (cp)a$, and so $a|(bc)$. ∎

12.15 a. For all s in S, $s \leq x$.

b. There is an s in S such that $s \geq x$.

c. There is an x with $ax \leq b$ and $x \geq 0$ such that $cx \leq u$.

d. There is an x with $ax \leq b$ and $x \geq 0$ such that $cx \geq u$.

e. For all x with $b \leq x \leq c$, $ax \geq u$.

f. For all x with $b \leq x \leq c$, $ax \leq u$.

12.17 ***Analysis of proof.*** The max/min method can be used to convert the conclusion into the equivalent statement:

B1: for all s in S, $s \geq t^*$.

The appearance of the quantifier "for all" in the backward process suggests using the choose method to choose

A1: an element s' in S,

for which it must be shown that

B2: $s' \geq t^*$.

The desired conclusion can be obtained by working forward and using specialization. Specifically, since S is a subset of T, it follows by definition that

A2: for all elements s in S, s is in T.

Specializing $A2$ to $s = s'$ which is in S (see $A1$), it follows that

A3: s' is in T.

Also, the hypothesis states that

A4: for all t in T, $t \geq t^*$.

Applying specialization to $A4$ with $t = s'$ which is in T (see $A3$), it follows that

A5: $s' \geq t^*$,

which is precisely $B2$, and so the proof is complete.

Proof. To reach the conclusion, let s' be in S. It will be shown that $s' \geq t^*$. By the hypothesis that S is a subset of T, it follows that s' is in T. But then the hypothesis ensures that $s' \geq t^*$. ∎

SOLUTIONS TO CHAPTER 13

13.1

a. Contrapositive or contradiction method, since the word "no" appears in the conclusion.

b. Induction method, since statement B is true for every integer $n \geq 4$.

c. Forward–backward method, since there is no apparent form to B.

d. Max/min method, since B has the word "maximum" in it.

e. Uniqueness method, since there is supposed to be one and only one line.

f. Contradiction or contrapositive method, since the word "no" is the first key word to appear in the conclusion.

g. Forward–backward method, since there is no apparent form to B.

h. Choose method, since the first quantifier in B (from the left) is "for every."

i. Construction method, since the first quantifier in B (from the left) is "there is."

13.3

a. Using the induction method one would first have to show that $4! > 4^2$. Then one would assume that $n! > n^2$ and $n \geq 4$, and try to show that $(n+1)! > (n+1)^2$.

b. Using the choose method, one would choose an integer n' for which $n' \geq 4$. One would then try to show that $(n')! > (n')^2$.

c. Converting the statement to the form "if ... then ..." one obtains "if n is an integer ≥ 4 then $n! > n^2$." One would therefore assume that n is an integer ≥ 4 and try to show that $n! > n^2$.

d. Using the contradiction method, one would assume that there is an integer $n \geq 4$ such that $n! \leq n^2$, and try to reach a contradiction.

GLOSSARY OF MATHEMATICAL SYMBOLS

Symbol	Meaning	Page
$\Rightarrow$	implies	4
$\Leftrightarrow$	if and only if	32
$\in$	is an element of	54
$\subseteq$	subset	55
ϕ	empty set	55
$\sim$	not	38
$\forall$	for all (for each, etc.)	56
$\exists$	there is (there are, etc.)	47
$\ni$	such that	47
$\wedge$	and	30
$\vee$	or	30
$\blacksquare$	Q.E.D. (which was to be demonstrated)	18

INDEX